MICHEL FOUCAULT

The Birth of the Clinic

Michel Foucault was born in Poitiers, France, in 1926. He lectured in universities throughout the world; served as the director at the Institut Français in Hamburg, Germany and at the Institut de Philosophie at the Faculté des Lettres in the University of Clermont-Ferrand, France; and wrote frequently for French newspapers and reviews. At the time of his death in 1984, he held a chair at France's most prestigious institution, the Collège de France.

Books *by* MICHEL FOUCAULT

*Madness and Civilization: A History of Insanity in the
Age of Reason*

The Birth of the Clinic: An Archaeology of Medical Perception

The Order of Things: An Archaeology of the Human Sciences

The Archaeology of Knowledge (and *The Discourse on Language*)

*I, Pierre Rivière, having slaughtered my mother, my sister, and
my brother....A Case of Parricide in the Nineteenth Century*

Discipline and Punish: The Birth of the Prison

*Herculine Barbin, Being the Recently Discovered Memoirs of a
Nineteenth-Century French Hermaphrodite*

The History of Sexuality, Volume 1: An Introduction

The Use of Pleasure: Volume 2 of The History of Sexuality

The Care of the Self: Volume 3 of The History of Sexuality

*Power/Knowledge: Selected Interviews and Other Writings,
1927-1977*

The Foucault Reader (edited by Paul Rabinow)

THE BIRTH OF THE CLINIC

THE BIRTH
OF THE
CLINIC
An Archaeology of Medical Perception

•

MICHEL FOUCAULT

Translated from the French by
A. M. Sheridan Smith

VINTAGE BOOKS
A Division of Random House, Inc.
New York

VINTAGE BOOKS EDITION, APRIL 1994

Library of Congress Cataloging in Publication Data
Foucault, Michel.
The birth of the clinic.
Reprint of the ed. published by Pantheon Books, New York, in series:
World of man.
Translation of Naissance de la clinique.
Bibliography: p.
1. Medicine—History. 2. Medicine—Philosophy. I. Title.
[R133.F6913 1975]   362.1'1   74-3389
ISBN 0-679-75334-6

Contents

Translator's Note

One of the characteristics of Foucault's language is his repeated use of certain key words. Many of these present no difficulty to the translator. Others, however, have no normal equivalent. In such cases, it is generally preferable to use a single unusual word rather than a number of familiar ones. When Foucault speaks of *la clinique*, he is thinking of both clinical medicine and the teaching hospital. So if one wishes to retain the unity of the concept, one is obliged to use the rather odd-sounding 'clinic'. Similarly, I have used the unusual 'gaze' for the common '*regard*', except in the book's subtitle, where I have made a concession to the unprepared reader.

Preface

This book is about space, about language, and about death; it is about the act of seeing, the gaze.

Towards the middle of the eighteenth century, Pomme treated and cured a hysteric by making her take 'baths, ten or twelve hours a day, for ten whole months'. At the end of this treatment for the dessication of the nervous system and the heat that sustained it, Pomme saw 'membranous tissues like pieces of damp parchment . . . peel away with some slight discomfort, and these were passed daily with the urine; the right ureter also peeled away and came out whole in the same way'. The same thing occurred with the intestines, which at another stage, 'peeled off their internal tunics, which we saw emerge from the rectum. The oesophagus, the arterial trachea, and the tongue also peeled in due course; and the patient had rejected different pieces either by vomiting or by expectoration' [1].

Less than a hundred years later, this is how a doctor observed an anatomical lesion of the brain and its enveloping membranes, the so-called 'false membranes' frequently found on patients suffering from 'chronic meningitis:'

> Their outer surface, which is next to the arachnoidian layer of the dura mater, adheres to this layer, sometimes very lightly, when they can be separated easily, sometimes very firmly and tightly, in which case it can be very difficult to detach them. Their internal surface is only contiguous with the arachnoid, and is in no way joined to it . . . . The false membranes are often transparent,

especially when they are very thin; but usually they are white, grey, or red in colour, and occasionally, yellow, brown, or black. This matter often displays different shades in different parts of the same membrane. The thickness of these accidental productions varies greatly; sometimes they are so tenuous that they might be compared to a spider's web. . . . The organization of the false membranes also displays a great many differences: the thin ones are buffy, like the albuminous skins of eggs, and have no distinctive structure of their own. Others, on one of their sides, often display traces of blood vessels crossing over one another in different directions and injected. They can often be reduced to layers placed one upon another, between which discoloured blood clots are frequently interposed [2].

Between Pomme, who carried the old myths of nervous pathology to their ultimate form, and Bayle, who described the encephalic lesions of general paralysis for an era from which we have not yet emerged, the difference is both tiny and total. For us, it is total, because each of Bayle's words, with its qualitative precision, directs our gaze into a world of constant visibility, while Pomme, lacking any perceptual base, speaks to us in the language of fantasy. But by what fundamental experience can we establish such an obvious difference below the level of our certainties, in that region from which they emerge? How can we be sure that an eighteenth-century doctor did not see what he saw, but that it needed several decades before the fantastic figures were dissipated to reveal, in the space they vacated, the shapes of things as they really are?

What occurred was not a 'psychoanalysis' of medical knowledge, nor any more or less spontaneous break with imaginary investments; 'positive' medicine is not a medicine that has made an 'objectal' choice in favour of objectivity itself. Not all the powers of a visionary space through which doctors and patients, physiologists and practitioners communicated (stretched and twisted nerves, burning dryness, hardened or burnt organs, the new birth of the body in the beneficent element of cool waters) have disappeared; it is, rather, as if they had been displaced, enclosed within the singularity of the patient, in that region of 'subjective symptoms' that—for the doctor—defines not the mode of knowledge, but the world of objects to be known. Far from being broken, the fantasy link between knowledge and pain is reinforced by a more complex means than the mere permeability of the imagination;

the presence of disease in the body, with its tensions and its burnings, the silent world of the entrails, the whole dark underside of the body lined with endless unseeing dreams, are challenged as to their objectivity by the reductive discourse of the doctor, as well as established as multiple objects meeting his positive gaze. The figures of pain are not conjured away by means of a body of neutralized knowledge; they have been redistributed in the space in which bodies and eyes meet. What has changed is the silent configuration in which language finds support: the relation of situation and attitude to what is speaking and what is spoken about.

From what moment, from what semantic or syntactical change, can one recognize that language has turned into rational discourse? What sharp line divides a description that depicts membranes as being like 'damp parchment' from that other equally qualitative, equally metaphorical description of them laid out over the tunic of the brain, like a film of egg whites? Do Bayle's 'white' and 'red' membranes possess greater value, solidity, and objectivity—in terms of scientific discourse—than the horny scales described by the doctors of the eighteenth century? A rather more meticulous gaze, a more measured verbal tread with a more secure footing upon things, a more delicate, though sometimes rather confused choice of adjective—are these not merely the proliferation, in medical language, of a style which, since the days of galenic medicine, has extended whole regions of description around the greyness of things and their shapes?

In order to determine the moment at which the mutation in discourse took place, we must look beyond its thematic content or its logical modalities to the region where 'things' and 'words' have not yet been separated, and where—at the most fundamental level of language—seeing and saying are still one. We must re-examine the original distribution of the visible and invisible insofar as it is linked with the division between what is stated and what remains unsaid: thus the articulation of medical language and its object will appear as a single figure. But if one poses no retrospective question, there can be no priority; only the spoken structure of the perceived—that *full* space in the *hollow* of which language assumes volume and size—may be brought up into the indifferent light of day. We must place ourselves, and remain once and for all, at the level of the fundamental *spatialization* and *verbalization* of the pathological, where the loquacious gaze with which the doctor observes the poisonous heart of things is born and communes with itself.

Modern medicine has fixed its own date of birth as being in the last years of the eighteenth century. Reflecting on its situation, it identifies the origin of its positivity with a return—over and above all theory—to the modest but effecting level of the perceived. In fact, this supposed empiricism is not based on a rediscovery of the absolute values of the visible, nor on the predetermined rejection of systems and all their chimeras, but on a reorganization of that manifest and secret space that opened up when a millennial gaze paused over men's sufferings. Nonetheless the rejuvenation of medical perception, the way colours and things came to life under the illuminating gaze of the first clinicians is no mere myth. At the beginning of the nineteenth century, doctors described what for centuries had remained below the threshold of the visible and the expressible, but this did not mean that, after over-indulging in speculation, they had begun to perceive once again, or that they listened to reason rather than to imagination; it meant that the relation between the visible and invisible—which is necessary to all concrete knowledge—changed its structure, revealing through gaze and language what had previously been below and beyond their domain. A new alliance was forged between words and things, enabling one *to see* and *to say*. Sometimes, indeed, the discourse was so completely 'naive' that it seems to belong to a more archaic level of rationality, as if it involved a return to the clear, innocent gaze of some earlier, golden age.

In 1764, J. F. Meckel set out to study the alterations brought about in the brain by certain disorders (apoplexy, mania, phthisis); he used the rational method of weighing equal volumes and comparing them to determine which parts of the brain had been dehydrated, which parts had been swollen, and by which diseases. Modern medicine has made hardly any use of this research. Brain pathology achieved its 'positive' form when Bichat, and above all Récamier and Lallemand, used the celebrated 'hammer, with a broad, thin end. If one proceeds with light taps, no concussion liable to cause disorders can result as the skull is full. It is better to begin from the rear, because, when only the occipital has to be broken, it is often so mobile that one misses one's aim. . . . In the case of very young children, the bones are too supple to be broken and too thin to be sawn; they have to be cut with strong scissors' [3]. The fruit is then opened up. From under the meticulously parted shell, a soft, greyish mass appears, wrapped in viscous, veined skins: a delicate, dingy-looking pulp within which—freed at last and exposed at last to the light of day—shines the seat of knowledge. The antisanal skill of the brain-breaker has replaced the scientific precision of the scales, and yet our science since Bichat identifies with the former; the precise, but immeasurable gesture that opens up the plenitude of concrete things, combined with the delicate network of their properties to the gaze, has produced a more scientific objectivity for us than instrumental arbitrations of quantity. Medical rationality plunges into the marvelous density of perception, offering the grain of things as the first face of truth, with their colours, their spots, their hardness, their adherence. The breadth of the experiment seems to be identified with the domain of the careful gaze, and of an empirical vigilance receptive only to the evidence of visible contents. The eye becomes the depositary and source of clarity; it has the power to bring a truth to light that it receives only to the extent that it has brought it to light; as it opens, the eye first opens the truth: a flexion that marks the transition from the world of classical clarity—from the 'enlightenment' —to the nineteenth century.

For Descartes and Malebranche, to see was to perceive (even in the most concrete kinds of experience, such as Descartes's practice of anatomy, or Malebranche's microscopic observations); but, without stripping perception of its sensitive body, it was a matter of rendering it transparent for the exercise of the mind: light, anterior to every gaze, was the element of ideality—the unassignable place of origin where things were adequate to their essence— and the form by which things reached it through the geometry of bodies; according to them, the act of seeing, having attained perfection, was absorbed back into the unbending, unending figure of light. At the end of the eighteenth century, however, seeing consists in leaving to experience its greatest corporal opacity; the solidity, the obscurity, the density of things closed in upon themselves, have powers of truth that they owe not to light, but to the slowness of the gaze that passes over them, around them, and gradually into them, bringing them nothing more than its own light. The residence of truth in the dark centre of things is linked, paradoxically, to this sovereign power of the empirical gaze that turns their darkness into light. All light has passed over into the thin flame of the eye, which now flickers around solid objects and, in so doing, establishes their place and form. Rational discourse is based less on the geometry of light than on the insistent, impenetrable density of the object, for prior to all knowledge, the source, the domain, and the boundaries of experience can be found in its dark presence. The gaze is passively linked to the primary passivity that dedicates it to the endless task of absorbing experience in its entirety, and of mastering it.

The task lay with this language of things, and perhaps with it alone, to authorize a knowledge of the individual that was not simply of a historic or aesthetic order. That the definition of the individual should be an endless labour was no longer an obstacle to an experience, which, by accepting its own limits, extended its task into the infinite. By acquiring the status of object, its particular quality, its impalpable colour, its unique, transitory form took on weight and solidity. No light could now dissolve them in ideal truths; but the gaze directed upon them would, in turn, awaken them and make them stand out against a background of objectivity. The gaze is no longer reductive, it is, rather, that which establishes the individual in his irreducible quality. And thus it becomes possible to organize a rational language around it. The *object* of discourse may equally well be a *subject*, without the figures of objectivity being in any way altered. It is this *formal* reorganization, *in depth*, rather than the abandonment of theories and old systems, that made *clinical experience* possible; it lifted the old Aristotelian prohibition: one could at last hold a scientifically structured discourse about an individual.

Our contemporaries see in this accession to the individual the establishment of a 'unique dialogue', the most concentrated formulation of an old medical humanism, as old as man's compassion. The mindless phenomenologies of understanding mingle the sand of their conceptual desert with this half-baked notion; the feebly eroticized vocabulary of 'encounter' and of the 'doctor/patient relationship' (*le couple médecin-malade*) exhausts itself in trying to communicate the pale powers of matrimonial fantasies to so much non-thought. Clinical experience—that opening up of the concrete individual, for the first time in Western history, to the language of rationality, that major event in the relationship of man to himself and of language to things—was soon taken as a simple, unconceptualized confrontation of a gaze and a face, or a glance and a silent body; a sort of contact prior to all discourse, free of the burdens of language, by which two living individuals are 'trapped' in a common, but non-reciprocal situation. Recently, in the interests of an open market, so-called 'liberal' medicine has revived the old rights of a clinic understood as a special contract, a tacit pact made between one man and another. This patient gaze has even been attributed with the power of assuming—with the calculated addition of reasoning (neither too much nor too little)—the general form of all scientific observation:

> In order to be able to offer each of our patients a course of treatment perfectly adapted to his illness and to himself, we try to obtain a complete, objective idea of his case; we gather together in a file of his own all the information we have about him. We 'observe' him in the same way that we observe the stars or a laboratory experiment [4].

Miracles are not so easy to come by: the mutation that made it possible—and which continues to do so every day—for the patient's 'bed' to become a field of scientific investigation and discourse is not the sudden explosive mixture of an old practice and an even older logic, or that of a body of knowledge and some strange, sensorial element of 'touch', 'glance', or 'flair'. Medicine made its appearance as a clinical science in conditions which define, together with its historical possibility, the domain of its experience and the structure of its rationality. They form its concrete a priori, which it is now possible to uncover, perhaps because a new experience of disease is coming into being that will make possible a historical and critical understanding of the old experience.

A detour is necessary here if we are to lay the foundations of our discourse on the birth of the clinic. It is a strange discourse, I admit, since it will be based neither on the present consciousness of clinicians, nor even on a repetition of what they once might have said.

It may well be that we belong to an age of criticism whose lack of a primary philosophy reminds us at every moment of its reign and its fatality: an age of intelligence that keeps us irremediably at a distance from an original language. For Kant, the possibility and necessity of a critique were linked, through certain scientific contents, to the fact that there is such a thing as knowledge. In our

time—and Nietzsche the philologist testifies to it—they are linked
to the fact that language exists and that, in the innumerable words
spoken by men—whether they are reasonable or senseless, demon-
strative or poetic—a meaning has taken shape that hangs over us,
leading us forward in our blindness, but awaiting in the darkness
for us to attain awareness before emerging into the light of day and
speaking. We are doomed historically to history, to the patient
construction of discourses about discourses, and to the task of hearing what has already been said.

But is it inevitable that we should know of no other function
for speech (*parole*) than that of commentary? *Commentary* questions discourse as to what it says and intended to say; it tries to
uncover that deeper meaning of speech that enables it to achieve
an identity with itself, supposedly nearer to its essential truth;
in other words, in stating what has been said, one has to re-state
what has never been said. In this activity known as commentary
which tries to transmit an old, unyielding discourse seemingly
silent to itself, into another, more prolix discourse that is both
more archaic and more contemporary—is concealed a strange attitude towards language: to comment is to admit by definition
an excess of the signified over the signifier; a necessary, unformu-
lated remainder of thought that language has left in the shade—
a remainder that is the very essence of that thought, driven outside
its secret—but to comment also presupposes that this unspoken
element slumbers within speech (*parole*), and that, by a super-
abundance proper to the signifier, one may, in questioning it, give
voice to a content that was not explicitly signified. By opening
up the possibility of commentary, this double plethora dooms us
to an endless task that nothing can limit: there is always a certain
amount of signified remaining that must be allowed to speak, while
the signifier is always offered to us in an abundance that questions
us, in spite of ourselves, as to what it 'means' (*veut dire*). Signifier
and signified thus assume a substantial autonomy that accords the
treasure of a virtual signification to each of them separately; one
may even exist without the other, and begin to speak of itself:
commentary resides in that supposed space. But at the same time,
it invents a complex link between them, a whole tangled web that
concerns the poetic values of expression: the signifier is not supposed to 'translate' without concealing, without leaving the signified
with an inexhaustible reserve; the signified is revealed only in the
visible, heavy world of a signifier that is itself burdened with a

meaning that it cannot control. Commentary rests on the postulate that speech (*parole*) is an act of 'translation', that it has the dangerous privilege images have of showing while concealing, and that it can be substituted for itself indefinitely in the open series of discursive repetitions; in short, it rests on a psychologistic interpretation of language that shows the stigmatas of its historical origin. This is an exegesis, which listens, through the prohibitions, the symbols, the concrete images, through the whole apparatus of Revelation, to the Word of God, ever secret, ever beyond itself. For years we have been commenting on the language of our culture from the very point where for centuries we had awaited in vain for the decision of the Word.

To speak about the thought of others, to try to say what they have said has, by tradition, been to analyse the signified. But must the things said, elsewhere and by others, be treated exclusively in accordance with the play of signifier and signified, as a series of themes present more or less implicitly to one another? Is it not possible to make a structural analysis of discourses that would evade the fate of commentary by supposing no remainder, nothing in excess of what has been said, but only the fact of its historical appearance? The facts of discourse would then have to be treated not as autonomous nuclei of multiple significations, but as events and functional segments gradually coming together to form a system. The meaning of a statement would be defined not by the treasure of intentions that it might contain, revealing and concealing it at the same time, but by the difference that articulates it upon the other real or possible statements, which are contemporary to it or to which it is opposed in the linear series of time. A systematic history of discourses would then become possible.

Until recently, the history of ideas was only aware of two methods: the first, aesthetic method involved analogy, with diffusion charted in time (geneses, filiations, kinships, influences) or on the surface of a given historical space (the spirit of a period, its *Weltanschauung*, its fundamental categories, the organization of its sociocultural world). The second, which was a psychological method, involved a denial of contents (this or that century was not as rationalistic, or irrationalistic as was said or believed), from which there has since developed a sort of 'psychoanalysis' of thought, the results of which can quite legitimately be reversed—the nucleus of the nucleus being always its opposite.

I should like to attempt here the analysis of a type of discourse—

that of medical experience—at a period when, before the great
discoveries of the nineteenth century, it had changed its materials
more than its systematic form. The clinic is both a new 'carving up'
of things and the principle of their verbalization in a form which
we have been accustomed to recognizing as the language of a
'positive science'.

To anyone wishing to draw up an inventory of its themes, the
idea of the clinic would undoubtedly seem to be imbued with rather
vague values; insipid figures would probably take shape, such as
the strange effect of disease on the patient, the diversity of individual temperaments, the probability of pathological evolution, the
need for sharp perception (the need to be constantly alert to the
slightest visible modalities), the empirical form—cumulative, and
endlessly open to medical knowledge—old, threadbare notions that
had been medicine's basic tools as far back as the Greeks. Nothing
in this ancient arsenal can designate clearly what took place at that
turning point in the eighteenth century, when the calling into
question of the old clinical theme 'produced'—if we are to believe
first appearances—an essential mutation in medical knowledge.
Nonetheless, considered on an over-all basis, the clinic appears—
in terms of the doctor's experience—as a new outline of the perceptible and statable: a new distribution of the discrete elements of
corporal space (for example, the isolation of *tissue*—a functional,
two-dimensional area—in contrast with the functioning mass of the
organ, constituting the paradox of an 'internal surface') a reorganization of the elements that make up the pathological phenomenon
(a grammar of signs has replaced a botany of symptoms), a definition of the linear series of morbid events (as opposed to the table
of nosological species), a welding of the disease onto the organism
(the disappearance of the general morbid entities that grouped
symptoms together in a single logical figure, and their replacement
by a local status that situates the being of the disease with its causes
and effects in a three-dimensional space). The appearance of the
clinic as a historical fact must be identified with the system of these
reorganizations. This new structure is indicated—but not, of course,
exhausted—by the minute but decisive change, whereby the question: 'What is the matter with you?', with which the eighteenth-
century dialogue between doctor and patient began (a dialogue
possessing its own grammar and style), was replaced by that other
question: 'Where does it hurt?', in which we recognize the operation of the clinic and the principle of its entire discourse. From

then on, the whole relationship of signifier to signified, at every level of medical experience, is redistributed: between the symptoms that signify and the disease that is signified, between the description and what is described, between the event and what it prognosticates, between the lesion and the pain that it indicates, etc. The clinic—constantly praised for its empiricism, the modesty of its attention, and the care with which it silently lets things surface to the observing gaze without disturbing them with discourse— owes its real importance to the fact that it is a reorganization in depth, not only of medical discourse, but of the very possibility of a discourse about disease. The *restraint* of clinical discourse (its rejection of theory, its abandonment of systems, its lack of a philosophy; all so proudly proclaimed by doctors) reflects the non-verbal conditions on the basis of which it can speak: the common structure that carves up and articulates what is seen and what is said.

The research that I am undertaking here therefore involves a project that is deliberately both historical and critical, in that it is concerned—outside all prescriptive intent—with determining the conditions of possibility of medical experience in modern times.

I should like to make it plain once and for all that this book has not been written in favour of one kind of medicine as against another kind of medicine, or against medicine and in favour of an absence of medicine. It is a structural study that sets out to disentangle the conditions of its history from the density of discourse, as do others of my works.

What counts in the things said by men is not so much what they may have thought or the extent to which these things represent their thoughts, as that which systematizes them from the outset, thus making them thereafter endlessly accessible to new discourses and open to the task of transforming them.

NOTES

[1] Pomme, *Traité des affections vaporeuses des deux sexes* (4th edn., Lyons, 1769, vol. I, pp. 60–5).

[2] A. L. J. Bayle, *Nouvelle doctrine des maladies mentales* (Paris, 1825, pp. 23–4).

[3] F. Lallemand, *Recherches anatomo-pathologiques sur l'encéphale* (Paris, 1820, introduction, p. vii, n.).

[4] J. -Ch. Sournia, *Logique et morale du diagnostic* (Paris, 1962, p. 19).

THE BIRTH OF THE CLINIC

1 · Spaces and Classes

For us, the human body defines, by natural right, the space of origin and of distribution of disease: a space whose lines, volumes, surfaces, and routes are laid down, in accordance with a now familiar geometry, by the anatomical atlas. But this order of the solid, visible body is only one way—in all likelihood neither the first, nor the most fundamental—in which one spatializes disease. There have been, and will be, other distributions of illness.

When will we be able to define the structures that determine, in the secret volume of the body, the course of allergic reactions? Has anyone ever drawn up the specific geometry of a virus diffusion in the thin layer of a segment of tissue? Is the law governing the spatialization of these phenomena to be found in a Euclidean anatomy? After all, one only has to remember that the old theory of sympathies spoke a vocabulary of correspondences, vicinities, and homologies, terms for which the perceived space of anatomy hardly offers a coherent lexicon. Every great thought in the field of pathology lays down a configuration for disease whose spatial requisites are not necessarily those of classical geometry.

The exact superposition of the 'body' of the disease and the body of the sick man is no more than a historical, temporary datum. Their encounter is self-evident only for us, or, rather, we are only just beginning to detach ourselves from it. The space of *configuration* of the disease and the space of *localization* of the illness in the body have been superimposed, in medical experience, for only a relatively short period of time—the period that coincides with

3

nineteenth-century medicine and the privileges accorded to pathological anatomy. This is the period that marks the suzerainty of the
gaze, since in the same perceptual field, following the same con-
tinuities or the same breaks, experience reads at a glance the visible
lesions of the organism and the coherence of pathological forms;
the illness is articulated exactly on the body, and its logical distribution is carried out at once in terms of anatomical masses. The
'glance' has simply to exercise its right of origin over truth.

But how did this supposedly natural, immemorial right come
about? How was this locus, in which disease indicated its presence,
able to determine in so sovereign a way the figure that groups its
elements together? Paradoxically, never was the space of configuration of disease more free, more independent of its space of localization than in classificatory medicine, that is to say, in that form of
medical thought that, historically, just preceded the anatomo-clinical method, and made it structurally possible.

'Never treat a disease without first being sure of its species,'
said Gilibert [1]. From the *Nosologie* of Sauvages (1761) to the
Nosographie of Pinel (1798), the classificatory rule dominates
medical theory and practice: it appears as the immanent logic of
morbid forms, the principle of their decipherment, and the semantic
rule of their definition: 'Pay no heed to those envious men who
would cast the shadow of contempt over the writings of the celebrated Sauvages. . . . Remember that of all the doctors who have
ever lived he is perhaps the only one to have subjected all our
dogmas to the infallible rules of healthy logic. Observe with what
care he defines his words, with what scrupulousness he circumscribes the definitions of each malady.' Before it is removed from
the density of the body, disease is given an organization, hierar-
chized into families, genera, and species. Apparently, this is no more
than a 'picture' that helps us to learn and to remember the prolifer-
ating domain of the diseases. But at a deeper level than this spatial
'metaphor', and in order to make it possible, classificatory medicine
presupposes a certain 'configuration' of disease: it has never been
formulated for itself, but one can define its essential requisites after
the event. Just as the genealogical tree, at a lower level than the
comparison that it involves and all its imaginary themes, presupposes
a space in which kinship is formalizable, the nosological picture
involves a figure of the diseases that is neither the chain of causes
and effects nor the chronological series of events nor its visible
trajectory in the human body.

This organization treats localization in the organism as a sub-
sidiary problem, but defines a fundamental system of relations involving envelopments, subordinations, divisions, resemblances. This
space involves: a 'vertical', in which the implications are drawn up
—fever, 'a successive struggle between cold and heat', may occur
in a single episode, or in several; these may follow without interruption or after an interval; this respite may not exceed twelve
hours, attain a whole day, last two whole days, or have a poorly
defined rhythm [2]; and a 'horizontal', in which the homologies
are transferred—in the two great subdivisions of the spasms are to
be found, in perfect symmetry, the 'partial tonics', the 'general
tonics', the 'partial clonics', and the 'general clonics' [3]; or again,
in the order of the discharges, what catarrh is to the throat, dysentery is to the intestines [4]; a deep space, anterior to all perceptions,
and governing them from afar; it is on the basis of this space, the
lines that it intersects, the masses that it distributes or hierarchizes,
that disease, emerging beneath our gaze, becomes embodied in a
living organism.

What are the principles of this primary configuration of disease?

1. The doctors of the eighteenth century identified it with
'historical', as opposed to philosophical, 'knowledge'. Knowledge
is historical that circumscribes pleurisy by its four phenomena:
fever, difficulty in breathing, coughing, and pains in the side.
Knowledge would be philosophical that called into question the
origin, the principle, the causes of the disease: cold, serous discharge, inflammation of the pleura. The distinction between the
historical and the philosophical is not the distinction between cause
and effect: Cullen based his classificatory system on the attribution
of related causes [5]; nor is the distinction between principle and
consequences, since Sydenham thought he was engaged in historical
research when studying 'the way in which nature produces and
sustains the different forms of diseases' [6]; nor even is it exactly
the difference between the visible and the hidden or conjectural,
for one sometimes has to track down a 'history' that is enclosed
upon itself and develops invisibly, like hectic fever in certain
phthisics: 'reefs caught under water' [7]. The historical embraces
whatever, *de facto* or *de jure*, sooner or later, directly or indirectly,
may be offered to the gaze. A cause that can be seen, a symptom
that is gradually discovered, a principle that can be deciphered
from its root do not belong to the order of 'philosophical' knowl-

edge, but to a 'very simple' knowledge, which 'must precede all others', and which situates the original form of medical experience. It is a question of defining a sort of fundamental area in which perspectives are levelled off, and in which shifts of level are aligned: an effect has the same status as its cause, the antecedent coincides with what follows it. In this homogeneous space series are broken and time abolished: a local inflammation is merely the ideal juxtaposition of its historical elements (redness, tumour, heat, pain) without their network of reciprocal determinations or their temporal intersection being involved.

Disease is perceived fundamentally in a space of projection without depth, of coincidence without development. There is only one plane and one moment. The form in which truth is originally shown is the surface in which relief is both manifested and abolished —the portrait: 'He who writes the history of diseases must . . . observe attentively the clear and natural phenomena of diseases, however uninteresting they may seem. In this he must imitate the painters who when they paint a portrait are careful to mark the smallest signs and natural things that are to be found on the face of the person they are painting' [8]. The first structure provided by classificatory medicine is the flat surface of perpetual simultaneity. Table and picture.

2. It is a space in which analogies define essences. The pictures resemble things, but they also resemble one another. The *distance* that separates one disease from another can be measured only by the *degree* of their *resemblance*, without reference to the logico-temporal divergence of genealogy. The disappearance of voluntary movements and reduced activity in the internal or external sense organs form the general outline that emerges beneath such particular forms as apoplexy, syncope, or paralysis. Within this great kinship, minor divergences are established: apoplexy robs one of the use of all the senses, and of all voluntary motility, but it spares the breathing and the functioning of the heart; paralysis affects only a locally assignable sector of the nervous system and motility; like apoplexy, syncope has a general effect, but it also interrupts respiratory movements [9]. The perspective distribution, which enables us to see in paralysis a symptom, in syncope an episode, and in apoplexy an organic and functional attack, does not exist for the classificatory gaze, which is sensitive only to surface divisions, in which vicinity is not defined by measurable distances but

by formal similarities. When they become dense enough, these similarities cross the threshold of mere kinship and accede to unity of essence. There is no fundamental difference between an apoplexy that suddenly suspends motility, and the chronic, evolutive forms that gradually invade the whole motor system: in that simultaneous space in which forms distributed by time come together and are superimposed, kinship folds back into identity. In a flat, homogeneous, non-measurable world, there is essential disease where there is a plethora of similarities.

3. The form of the similarity uncovers the rational order of the diseases. When one perceives a resemblance, one does not simply lay down a system of convenient, relative 'mappings'; one begins to read off the intelligible ordering of the diseases. The veil is lifted from the principle of their creation; this is the general order of nature. As in the case of plants or animals, the action of disease is fundamentally specific: 'The supreme Being is not subjected to less certain laws in producing diseases or in maturing morbific humours, than in growing plants and animals. . . . He who observes attentively the order, the time, the hour at which the attack of quart fever begins, the phenomena of shivering, of heat, in a word all the symptoms proper to it, will have as many reasons to believe that this disease is a species as he has to believe that a plant constitutes a species because it grows, flowers, and dies always in the same way' [10].

This botanical model has a double importance for medical thought. First, it made it possible to turn the principle of the analogy of forms into the law of the production of essences; and, secondly, it allowed the perceptual attention of the doctor—which, here and there, discovers and relates—to communicate with the ontological order—which organizes from the inside, prior to all manifestation—the world of disease. The order of disease is simply a 'carbon copy' of the world of life; the same structures govern each, the same forms of division, the same ordering. The rationality of life is identical with the rationality of that which threatens it. Their relationship is not one of nature and counter-nature; but, in a natural order common to both, they fit into one another, one superimposed upon the other. In disease, one *recognizes* (*reconnaît*) life because it is on the law of life that *knowledge* (*connaissance*) of the disease is also based.

4. We are dealing with species that are both natural and ideal.

Natural, because it is in them that diseases state their essential truths; ideal insofar as they are never experienced unchanged and undisturbed.

The first disturbance is introduced with and by disease itself. To the pure nosological essence, which fixes and exhausts its place in the order of the species without residue, the patient adds, in the form of so many disturbances, his predispositions, his age, his way of life, and a whole series of events that, in relation to the essential nucleus, appear as accidents. In order to know the truth of the pathological fact, the doctor must abstract the patient: 'He who describes a disease must take care to distinguish the symptoms that necessarily accompany it, and which are proper to it, from those that are only accidental and fortuitous, such as those that depend on the temperament and age of the patient' [11]. Paradoxically, in relation to that which he is suffering from, the patient is only an external fact; the medical reading must take him into account only to place him in parentheses. Of course, the doctor must know 'the internal structure of our bodies'; but only in order to subtract it, and to free to the doctor's gaze 'the nature and combination of symptoms, crises, and other circumstances that accompany diseases' [12]. It is not the pathological that functions, in relation to life, as a *counter-nature*, but the patient in relation to the disease itself.

And not only the patient; the doctor, too. His intervention is an act of violence if it is not subjected strictly to the ideal ordering of nosology: 'The knowledge of diseases is the doctor's compass; the success of the cure depends on an exact knowledge of the disease'; the doctor's gaze is directed initially not towards that concrete body, that visible whole, that positive plenitude that faces him—the patient—but towards intervals in nature, lacunae, distances, in which there appear, like negatives, 'the signs that differentiate one disease from another, the true from the false, the legitimate from the bastard, the malign from the benign' [13]. It is a grid that catches the real patient and holds back any therapeutic indiscretion. If, for polemical reasons, the remedy is administered too early, it contradicts and blurs the essence of the disease; it prevents the disease from acceding to its true nature, and, by making it irregular, makes it untreatable. In the period of invasion, the doctor must hold his breath, for 'the beginnings of disease reveal its class, its genus, and its species'; when the symptoms increase and become more marked, it is enough 'to diminish their violence and reduce the pains'; when the disease has settled in, one must 'follow step by step the paths followed by nature', strengthening it if it is too weak, diminishing it if it strives too vigorously to destroy what resists it' [14].

In the rational space of disease, doctors and patients do not occupy a place as of right; they are tolerated as disturbances that can hardly be avoided: the paradoxical role of medicine consists, above all, in neutralizing them, in maintaining the maximum difference between them, so that, in the void that appears between them, the ideal configuration of the disease becomes a concrete, free form, totalized at last in a motionless, simultaneous picture, lacking both density and secrecy, where recognition opens of itself onto the order of essences.

Classificatory thought gives itself an essential space, which it proceeds to efface at each moment. Disease exists only in that space, since that space constitutes it as nature; and yet it always appears rather out of phase in relation to that space, because it is manifested in a real patient, beneath the observing eye of a forearmed doctor. The fine two-dimensional space of the portrait is both the origin and the final result: that which makes possible, at the outset, a rational, well-founded body of medical knowledge, and that towards which it must constantly proceed through that which conceals it. One of the tasks of medicine, therefore, is to rejoin its own condition, but by a path in which it must efface each of its steps, because it attains its aim in a gradual neutralization of itself. The condition of its truth is the necessity that blurs its outlines. Hence the strange character of the medical gaze; it is caught up in an endless reciprocity. It is directed upon that which is visible in the disease—but on the basis of the patient, who hides this visible element even as he shows it; consequently, in order to know, he must recognize, while already being in possession of the knowledge that will lend support to his recognition. And, as it moves forward, this gaze is really retreating, since it reaches the truth of the disease only by allowing it to win the struggle and to fulfill, in all its phenomena, its true nature.

Disease, which can be mapped out on the picture, becomes apparent in the body. There it meets a space with a quite different configuration: the concrete space of perception. Its laws define the visible forms assumed by disease in a sick organism: the way in which disease is distributed in the organism, manifests its presence there, progresses by altering solids, movements, or functions, causes lesions that become visible under autopsy, triggers off, at one point or another, the interplay of symptoms, causes reactions, and thus moves towards a fatal, and for it favourable, outcome. We are dealing here with those complex, derived figures by means of which the essence of the disease, with its structure of a picture, is articulated upon the thick, dense volume of the organism and becomes *embodied* within it.

How can the flat, homogeneous, homological space of classes become visible in a geographical system of masses differentiated by their volume and distance? How can a disease, defined by its *place* in a family, be characterized by its *seat* in an organism? This is the problem that might be called the *secondary spatialization* of the pathological.

For classificatory medicine, presence in an organ is never absolutely necessary to define a disease: this disease may travel from one point of localization to another, reach other bodily surfaces, while remaining identical in nature. The space of the body and the space of the disease possess enough latitude to slide away from one another. The same, single spasmodic malady may move from the lower part of the abdomen, where it may cause dyspepsia, visceral congestion, interruption of the menstrual or haemorrhoidal flow, towards the chest, with breathlessness, palpitations, the feeling of a lump in the throat, coughing, and finally reach the head, causing epileptic convulsions, syncopes, or sleepiness [15]. These movements, which are accompanied by symptomatic changes, may occur in time in a single individual; they may also be found by examining a series of individuals with different link points: in its visceral form, spasm is encountered, above all, in lymphatic subjects, while in its cerebral form it is encountered more among sanguine temperaments. But in any case, the essential pathological configuration is not altered. The organs are the concrete supports of the disease; they never constitute its indispensable conditions. The system of points that defines the relation of the disease to the organism is neither constant nor necessary. They do not possess a common, previously defined space.

In this corporal space in which it circulates freely, disease undergoes metastases and metamorphoses. Nothing confines it to a particular course. A nosebleed may become haemoptysis (spitting of blood) or cerebral haemorrhage; the only thing that must remain is the specific form of blood discharge. This is why the medicine of spaces has, throughout its history, been linked to the doctrine of sympathies—each notion being compelled to reinforce the other for the correct balance of the system. Sympathetic communication through the organism is sometimes carried out by a locally assignable relay (the diaphragm for spasms, the stomach for the discharge of humour); sometimes by a whole system of diffusion that radiates through the body (the nervous system for pains and convulsions, the vascular system for inflammations); in other cases, by means of a simple functional correspondence (a suppression of the excretions is communicated from the intestines to the kidneys, and from these to the skin); lastly, by means of an adjustment of the nervous system from one region to another (lumbar pains in the hydrocele). But the anatomical redistribution of the disease, whether through correspondence, diffusion, or relay, does not alter its essential structure; sympathy operates the interplay between the space of localization and the space of configuration; it defines their reciprocal freedom and the boundaries of that freedom.

Or, rather, threshold, not boundary. For beyond the sympathetic transference of the structural homology that it authorizes, a relation may be set up between one disease and another that is a relation of causality, but not of kinship. By virtue of its own creative force, one pathological form may engender another that is very far removed in the nosological picture. Hence the complications; hence the mixed forms; hence certain regular, or at least frequent, successions, as that between mania and paralysis. Haslam knew of delirious patients whose 'speech is disturbed, whose mouths are twisted, whose arms and legs are deprived of voluntary movement, whose memory is weakened', and who, generally speaking, 'have no awareness of their position' [16]. Overlapping of the symptoms or simultaneity of their extreme forms are not enough to constitute a single disease; the distance between verbal excitation and motor paralysis in the table of morbid kinships prevents a chronological proximity from deciding on a unity. Hence the idea of a causality that moves by virtue of a slight time-lag; sometimes the onset of mania appears first, sometimes the motor signs introduce the whole set of symptoms. 'The paralytic affections are a much more frequent cause of madness than is thought; and they are also a very common effect of mania.' No sympathetic translation can cross this gap between the species; and the solidarity of the symptoms in the organism are not enough to constitute a unity that clashes with the essences. There is, therefore, an inter-nosological causality, whose role is the contrary of sympathy: sympathy preserves the fundamental form by ranging over time and space; causality dissociates the simultaneities and intersections in order to maintain the essential purities.

In this pathology, time plays a limited role. It is admitted that a disease may last, and that its various episodes may appear in turn; ever since Hippocrates doctors have calculated the critical days of a disease, and known the significant values of the arterial pulsations: 'When the rebounding pulse appears at each thirtieth pulsation, or thereabouts, the haemorrhage occurs four days later, more or less; when it occurs at every sixteenth pulsation, the haemorrhage will occur in three days' time. . . . Lastly, when it recurs every fourth, third, second pulsation, or when it is continual, one must expect the haemorrhage within twenty-four hours' [17]. But this numerically fixed duration is part of the essential structure of disease, just as chronic catarrh becomes, after a period of time, phthisic fever. There is no process of evolution in which duration introduces new events of itself and at its own insistence; time is integrated as a nosological constant, not as an organic variable. The time of the body does not affect, and still less determines, the time of the disease.

What communicates the essential 'body' of the disease to the real body of the patient are not, therefore, the points of localization, nor the effects of duration, but, rather, the quality. In one of the experiments described before the Prussian Royal Academy in 1764, Meckel explains how he observed the alteration in the brain during different diseases. When he carried out an autopsy, he removed from the brain small cubes of equal volume ('6 lines in each direction') in different places in the cerebral mass: he compared these extractions with each other, and with similar cubes taken from other corpses. The instrument used for this comparison were weighing scales; in phthisis, a disease involving exhaustion, the specific weight of the brain was found to be relatively lower than in the case of apoplexy, a disease involving discharge (1 dr 3 ¾ gr as against 1 dr 6 or 7 gr); whereas in the case of a normal subject who had died naturally the average weight was 1 dr 5 gr. These weights may vary according to the part of the brain from which the samples have been extracted: in phthisis it is, above all, the cerebellum that is light; in apoplexy the central areas are heavy [18]. Between the disease and the organism, then, there are connexion points that are situated according to a regional principle; but these are only sectors in which the disease secretes or transposes its specific qualities: the brains of maniacs are light, dry, and friable because mania is a lively, hot, explosive disease; those of phthisics are exhausted and languishing, inert, anaemic, because phthisis belongs to the general class of the haemorrhages. The set of qualities characterizing a disease is laid down in an organ, which then serves as a support for the symptoms. The disease and the body communicate only through the non-spatial element of quality.

It is understandable, then, that medicine should turn away from what Sauvages called a 'mathematical' form of knowledge: 'Knowing quantities and being able to measure them, being able, for example, to determine the force and speed of the pulse, the degree of heat, the intensity of pain, the violence of the cough, and other such symptoms' [19]. Meckel measured, not to obtain knowledge of mathematical form, but to gauge the intensity of the pathological quality that constituted the disease. No measurable mechanics of the body can, in its physical or mathematical particularities, account for a pathological phenomenon; convulsions may be due to a dehydration and contraction of the nervous system—and this is certainly a phenomenon of a mechanical order; but it is a mechanics of interlinked qualities, articulated movements, upheavals that are triggered off in series, not a mechanics of quantifiable segments. It may involve a mechanism, but it cannot belong to the order of Mechanics as such. 'Physicians must confine themselves to knowing the forces of medicines and diseases by means of their operations; they must observe them with care and strive to know their laws, and be tireless in the search for physical causes' [20]. A true mathematization of disease would imply a common, homogeneous space, with organic figures and a nosological ordering.

On the contrary, their shift implies a qualitative gaze; in order to grasp the disease, one must look at those parts where there is dryness, ardour, excitation, and where there is humidity, discharge, debility. How can one distinguish, beneath the same fever, the same coughing, the same tiredness, pleurisy of the phthisis, if one does not recognize here a dry inflammation of the lungs, and there a serous discharge? How can one distinguish, if not by their quality, the convulsions of an epileptic suffering from cerebral inflammation, and those of a hypochondriac suffering from congestion of the viscera? A subtle perception of qualities, a perception of the differences between one case and another, a delicate perception of variants—a whole hermeneutics of the pathological fact, based on modulated, coloured experience, is required; one should measure variations, balances, excesses, and defects.

> The human body is made up of vessels and fluids; . . . when the vessels and fibres have neither too much nor too little tone, when the fluids have just the right consistency, when they have neither too much nor too little movement, man is in a state of health; if the movement . . . is too strong, the solids harden and the fluids thicken; if it is too weak, the fibre slackens and the blood becomes thinner [21].

And the medical gaze, open to these fine qualities, necessarily becomes attentive to all their modulations; the decipherment of disease in its specific characteristics is based on a subtle form of perception that must take account of each particular equilibrium. But in what does this particularity consist? It is not that of an organism in which pathological process and reactions are linked together in a unique way to form a 'case'. We are dealing, rather, with qualitative varieties of the illness, to which are added the varieties that may be presented by the temperaments, thus modulating the qualitative varieties in the second stage. What classificatory medicine calls 'particular histories' are the effects of multiplication caused by the qualitative variations (owing to the temperaments) of the essential qualities that characterize illnesses. The individual patient finds himself at the point at which the result of this multiplication appears.

Hence his paradoxical position. If one wishes to know the illness from which he is suffering, one must subtract the individual, with his particular qualities: 'The author of nature,' said Zimmermann, 'has fixed the course of most diseases through immutable laws that one soon discovers if the course of the disease is not interrupted or disturbed by the patient' [22]; at this level the individual was merely a negative element, the accident of the disease, which, for it and in it, is most alien to its essence. But the individual now reappears as the positive, ineffaceable support of all these qualitative phenomena, which articulate upon the organism the fundamental ordering of the disease; it is the local, sensible presence of this order—a segment of enigmatic space that unites the nosological plane of kinships to the anatomic volume of vicinities. The patient is a geometrically impossible spatial synthesis, but for that very reason unique, central, and irreplaceable: an order that has become density in a set of qualifying modulations. And the same Zimmermann, who recognized in the patient only the negative of the disease, is 'sometimes tempted', contrary to Sydenham's general descriptions, 'to admit only of particular histories. However simple nature may be as a whole, it is nevertheless varied in its parts; consequently, we must try to know it both as a whole and in its parts' [23]. The medicine of species becomes engaged in a renewed attention to the individual —an ever-more impatient attention, ever less able to tolerate the general forms of perception and the hasty inspection of essences.

'Every morning a certain Aesculapius has fifty or sixty patients in his waiting room; he listens to the complaints of each, arranges them into four lines, prescribes a bleeding for the first, a purge for the second, a clyster for the third, and a change of air for the fourth [24]. This is not medicine; the same is true of hospital practice, which kills the capacity for observation and stifles the talents of the observer by the sheer number of things to observe. Medical perception must be directed neither to series nor to groups; it must be structured as a look through 'a magnifying glass, which, when applied to different parts of an object, makes one notice other parts that one would not otherwise perceive' [25], thus initiating the endless task of understanding the individual. At this point, one is brought back to the theme of the portrait referred to above, but this time treated in reverse. The patient is the rediscovered portrait of the disease; he is the disease itself, with shadow and relief, modulations, nuances, depth; and when describing the disease the doctor must strive to restore this living density: 'One must render the patient's own infirmities, his own pains, his own gestures, his own posture, his own terms, and his own complaints' [26].

Through the play of primary spatialization, the medicine of species situated the disease in an area of homologies in which the individual could receive no positive status; in secondary spatialization, on the other hand, it required an acute perception of the individual, freed from collective medical structures, free of any group gaze and of hospital experience itself. Doctor and patient are caught up in an ever-greater proximity, bound together, the doctor by an ever-more attentive, more insistent, more penetrating gaze, the

patient by all the silent, irreplaceable qualities that, in him, betray—
that is, reveal and conceal—the clearly ordered forms of the disease.
Between the nosological characters and terminal features to be
read on the patient's face, the qualities have roamed freely over
the body. The medical gaze need hardly dwell on this body for
long, at least in its densities and functioning.

Let us call tertiary spatialization all the gestures by which, in a
given society, a disease is circumscribed, medically invested, isolated,
divided up into closed, privileged regions, or distributed throughout
cure centres, arranged in the most favorable way. Tertiary is not
intended to imply a derivative, less essential structure than the preceding ones; it brings into play a system of options that reveals
the way in which a group, in order to protect itself, practises ex-
clusions, establishes the forms of assistance, and reacts to poverty
and to the fear of death. But to a greater extent than the other forms
of spatialization, it is the locus of various dialectics: heterogeneous
figures, time lags, political struggles, demands and utopias, economic
constraints, social confrontations. In it, a whole corpus of medical
practices and institutions confronts the primary and secondary
spatializations with forms of a social space whose genesis, structure,
and laws are of a different nature. And yet, or, rather, for this very
reason, it is the point of origin of the most radical questionings.
It so happened that it was on the basis of this tertiary spatialization
that the whole of medical experience was overturned and defined
for its most concrete perceptions, new dimensions, and a new foundation.

In the medicine of species, disease has, as a birthright, forms and
seasons that are alien to the space of societies. There is a 'savage'
nature of disease that is both its true nature and its most obedient
course: alone, free of intervention, without medical artifice, it
reveals the ordered, almost vegetal nervure of its essence. But the
more complex the social space in which it is situated becomes, the
more *denatured* it becomes. Before the advent of civilization, people
had only the simplest, most necessary diseases. Peasants and workers
still remain close to the basic nosological table; the simplicity of
their lives allows it to show through in its reasonable order: they
have none of those variable, complex, intermingled nervous ills,
but down-to-earth apoplexies, or uncomplicated attacks of mania
[27]. As one improves one's conditions of life, and as the social
network tightens its grip around individuals, 'health seems to

diminish by degrees'; diseases become diversified, and combine with one another; 'their number is already great in the superior order of the bourgeois; . . . it is as great as possible in people of quality' [28].

Like civilization, the hospital is an artificial locus in which the transplanted disease runs the risk of losing its essential identity. It comes up against a form of complication that doctors call prison or hospital fever: muscular asthenia, dry or coated tongue, livid face, sticky skin, diarrhoea, pale urine, difficulty in breathing, death on the eighth or eleventh day, or on the thirteenth at the latest [29]. More generally, contact with other diseases, in this unkempt garden where the species cross-breed, alters the proper nature of the disease and makes it more difficult to decipher; and how in this necessary proximity can one 'correct the malign effluvium that exudes from the bodies of the sick, from gangrenous limbs, decayed bones, contagious ulcers, and putrid fevers'? [30] And, in any case, can one efface the unfortunate impression that the sight of these places, which for many are nothing more than 'temples of death', will have on a sick man or woman, removed from the familiar surroundings of his home and family? This loneliness in a crowd, this despair disturb, with the healthy reactions of the organism, the natural course of the disease; it would require a very skilful hospital doctor 'to avoid the danger of the false experience that seems to result from the artificial diseases to which he devotes himself in the hospitals. In fact, no hospital disease is a pure disease' [31].

The natural locus of disease is the natural locus of life—the family: gentle, spontaneous care, expressive of love and a common desire for a cure, assists nature in its struggle against the illness, and allows the illness itself to attain its own truth. The hospital doctor sees only distorted, altered diseases, a whole teratology of the pathological; the family doctor 'soon acquires true experience based on the natural phenomena of all species of disease' [32]. This family medicine must necessarily be respectful: 'Observe the sick, assist nature without violating it, and wait, admitting in all modesty that much knowledge is still lacking' [33]. Thus, on the subject of the pathology of species, there is a revival of the old dispute between active medicine and expectant medicine [34]. The nosologists of necessity favoured the latter, and one of these, Vitet, in a classification containing over two thousand species, and bearing the title *Médecine expectante*, invariably prescribes quina to help nature follow its natural course [35].

The medicine of species implies, therefore, a free spatialization for the disease, with no privileged region, no constraint imposed by hospital conditions—a sort of spontaneous division in the setting of its birth and development that must function as the paradoxical and natural locus of its own abolition. At the place in which it appears, it is obliged, by the same movement, to disappear. It must not be fixed in a medically prepared domain, but be allowed, in the positive sense of the term, to 'vegetate' in its original soil: the family, a social space conceived in its most natural, most primitive, most morally secure form, both enclosed upon itself and entirely transparent, where the illness is left to itself. Now, this structure coincides exactly with the way in which, in political thought, the problem of assistance is reflected.

The criticism levelled at hospital foundations was a commonplace of eighteenth-century economic analysis. The funds on which they are based are, of course, inalienable: they are the perpetual due of the poor. But poverty is not perpetual; needs change, and assistance must be given to those provinces and towns that need it. To do so would not be to contravene the wishes of the donors, but on the contrary to give them back their true form; their 'principal aim was to serve the public, to relieve the State; without departing from the intention of the founders, and even in conformity with their views, one must regard as a common mass all the funds donated to the hospitals' [36]. The single, sacrosanct foundation must be dissolved in favor of a generalized system of assistance, of which society is both the sole administrator and the undifferentiated beneficiary. Moreover, it is an error in economics to base assistance on an immobilization of capital—that is to say, on an impoverishment of the nation, which, in turn, brings with it the need for new foundations; hence, at worst, a stifling of activity. Assistance should be related neither to productive wealth (capital), nor to the wealth produced (profits, which are always capitalizable), but to the very principle that produces wealth: work. It is by giving the poor work that one will help the poor without impoverishing the nation [37].

The sick man is no doubt incapable of working, but if he is placed in a hospital he becomes a double burden for society: the assistance that he is given relates only to himself, and his family is, in turn, left exposed to poverty and disease. The hospital, which creates disease by means of the enclosed, pestilential domain that it constitutes, creates further disease in the social space in which it is placed. This separation, intended to protect, communicates disease and multiplies it to infinity. Inversely, if it is left in the free field of its birth and development, it will never be more than itself—as it appeared, so will it be extinguished—and the assistance that is given in the home will make up for the poverty that the disease has caused. The care spontaneously given by family and friends will cost nobody anything; and the financial assistance given to the sick man will be to the advantage of the family: 'someone will have to eat the meat from which his broth is made; and in heating his tisane, it costs no more to warm his children as well' [38]. The chain of one disease engendering another, and that of the perpetual impoverishment of poverty, is thus broken when one gives up trying to create for the sick a differentiated, distinct space, which results, in an ambiguous but clumsy way, in both the protection and the preservation of disease.

Independently of their justifications, the thought structure of the economists and that of the classificatory doctors coincide in broad terms: the space in which disease is isolated and reaches fulfilment is an absolutely open space, without either division or a privileged, fixed figure, reduced solely to the plane of visible manifestations; a homogeneous space in which no intervention is authorized except that of a gaze which is effaced as it alights, and of assistance whose sole value is its transitory compensation—a space with no other morphology than that of the resemblances perceived from one individual to another, and of the treatment administered by private medicine to a private patient.

But, by being carried to its conclusion in this way, the structure is inverted. Is a medical experience, diluted in the free space of a society reduced to the single, nodal, and necessary figure of the family, not bound up with the very structure of society? Does it not involve, because of the special attention that it pays to the individual, a generalized vigilance that by extension applies to the group as a whole? It would be necessary to conceive of a medicine sufficiently bound up with the state for it to be able, with the co-operation of the state, to carry out a constant, general, but differentiated policy of assistance; medicine becomes a task for the nation. (Menuret in the early days of the French Revolution dreamt of a system of free medical care administered by doctors who would be paid by the government out of the income from former church property [39].) In this way a certain supervision would be exercised over the doctors themselves; abuses would be prevented and quacks forbidden to practise, and, by means of an organized, healthy, rational medicine, home care would prevent the patient's becoming a victim of medicine and avoid exposure to contagion of the patient's family. Good medicine would be given status and legal protection by the state; and it would be the task of the state 'to make sure that a true art of curing does exist' [40]. The medicine of individual perception, of family assistance, of home care can be based only on a collectively controlled structure, or on one that is integrated into the social space in its entirety. At this point, a quite new form, virtually unknown in the eighteenth century, of institutional spatialization of disease, makes its appearance. The medicine of spaces disappears.

NOTES

[1] Gilibert, *L'anarchie médicinale* (Neuchâtel, 1772, vol. I, p. 198).
[2] F. Boissier de Sauvages, *Nosologie méthodique* (Lyons, 1772, vol. II).
[3] *Ibid.*, vol. III.
[4] W. Cullen, *Institutions de médecine pratique* (Fr. trans., Paris 1785, vol. II, pp. 39-60).
[5] W. Cullen, *Institutions de médecine pratique* (Fr. trans., Paris, 1785, 2 vols.).
[6] Th. Sydenham, *Médecine pratique* (Fr. trans. Jault, Paris, 1784, p. 390).
[7] *Ibid.*
[8] Th. Sydenham, quoted by Sauvages, *op. cit.*, vol. I, p. 88.
[9] W. Cullen, *op. cit.*, vol. II, p. 86.
[10] Sydenham, quoted by Sauvages, *op. cit.*, vol. I, pp. 124-5.
[11] *Ibid.*
[12] Clifton, *État de la médecine ancienne et moderne* (Fr. trans., Paris, 1742, p. 213).
[13] Frier, *Guide pour la conservation de l'homme* (Grenoble, 1789, p. 113).
[14] T. Guindant, *La nature opprimée par la médecine moderne* (Paris, 1768, pp. 10-11).
[15] *L'Encyclopédie*, article 'Spasme'.
[16] J. Haslam, *Observations on Madness* (London, 1798, p. 259).
[17] Fr. Solano de Luques, *Observations nouvelles et extraordinaires sur la prédiction des crises*, enlarged by several new cases by Nihell (Fr. trans., Paris, 1748, p. 2).
[18] Account in *Gazette salutaire*, vol. XXI, 2 August 1764.
[19] Sauvages, *op. cit.*, vol. I, pp. 91-2.
[20] Tissot, *Avis aux gens de lettres sur leur santé* (Lausanne, 1767, p. 28).
[21] *Ibid.*, p. 28.
[22] Zimmermann, *Traité de l'expérience* (Fr. trans., Paris, 1800, vol. I, p. 122).
[23] *Ibid.*, p. 184.

[24] *Ibid.*, p. 187.

[25] *Ibid.*, p. 127.

[26] *Ibid.*, p. 178.

[27] Tissot, *Traité des nerfs et de leurs maladies* (Paris, 1778–1780, vol. II, pp. 432–44).

[28] Tissot, *Essai sur la santé des gens du monde* (Lausanne, 1770, pp. 8–12).

[29] Tenon, *Mémoires sur les hôpitaux* (Paris, 1788, p. 451).

[30] Percival, 'Lettre à M. Aikin', in J. Aikin, *Observations sur les hôpitaux* (Fr. trans., Paris, 1777, p. 113).

[31] Dupont de Nemours, *Idées sur les secours à donner* (Paris, 1786, pp. 24–5).

[32] *Ibid.*

[33] Moscati, *De l'emploi des systèmes dans la médecine pratique* (Fr. trans., Strasbourg, Year VII, pp. 26–7).

[34] Cf. Vicq d'Azyr, *Remarques sur la médecine agissante* (Paris, 1786).

[35] Vitet, *La médecine expectante* (Paris, 1806, 6 vols.).

[36] Chamousset (C.H.P.), 'Plan général pour l'administration des hôpitaux', *Vues d'un citoyen* (Paris, 1757, vol. II).

[37] Turgot, article 'Fondation', in *L'Encyclopédie*.

[38] Dupont de Nemours, *op. cit.*, pp. 14–30.

[39] J.-J. Menuret, *Essai sur les moyens de former de bons médecins* (Paris, 1791).

[40] Jadelot, *Adresse à Nos Seigneurs de l'Assemblée Nationale sur la nécessité et le moyen de perfectionner l'enseignement de la médecine* (Nancy, 1790, p. 7).

2 · A Political Consciousness

Compared with the medicine of species, the notions of constitution, endemic disease, and epidemic were of only marginal importance in the eighteenth century.

But we must return to Sydenham and to the ambiguity of what he has to teach us: in addition to being the initiator of classificatory thought, he defined what might be a historical and geographical consciousness of disease. Sydenham's 'constitution' is not an autonomous nature, but the complex—a kind of temporary node—of a set of natural events: qualities of soil, climate, seasons, rain, drought, centres of pestilence, famine; and when all these factors do not account for phenomena, there remains no clear species in the garden of disease, but an obscure nucleus, buried in the earth: 'Variae sunt semper annorum constitutiones quae neque calori neque frigori non sicco humidove ortum suum debent, sed ab occulta potius inexplicabili quadam alternatione in ipsis terrae visceribus pendent' [1]. The constitutions hardly have symptoms of their own; they define, by displacements of accent, unexpected groups of signs, phenomena of a more intense or weaker kind: fevers may be violent and dry, catarrhs and serous discharges more frequent; during a long, hot summer, visceral congestion is more common and more tenacious than usual. Of London, between July and September 1661, Sydenham says: 'Aegri paroxysmus atrocior, lingua magis nigra siccaque, extra paroxysmum aporexia obscurio, virium et appetitus prostratio major, major item ad paroxysmum proclinitas, omnia summatim accidentia immaniora, ipseque morbus quam pro more Febrium

22

intermittentium funestior' [2]. The constitution is not related to a specific absolute of which it is the more or less modified manifestation: it is perceived solely in the relativity of differences—by a gaze that is in some sense diacritical.

Not every constitution is an epidemic; but an epidemic is a finer-grained constitution, with more constant, more homogeneous phenomena. There has been, and still is, a great deal of discussion as to whether the doctors of the eighteenth century had grasped its contagious character, and whether they had posed the problem of the agent of their transmission. An idle question, and one that remains alien, or at least derivative, in relation to the fundamental structure: an epidemic is more than a particular form of a disease. In the eighteenth century, it was an autonomous, coherent, and adequate evaluation of disease: 'One calls epidemic diseases all those that attack, at the same time and with unalterable characteristics, a large number of persons' [3]. There is no difference is nature or species, therefore, between an individual disease and an epidemic phenomenon; it is enough that a sporadic malady be reproduced a number of times for it to constitute an epidemic. It is a purely mathematical problem of the threshold: the sporadic disease is merely a submarginal epidemic. The perception involved is no longer essential and ordinal, as in the medicine of species, but quantitative and cardinal.

The basis of this perception is not a specific type, but a nucleus of circumstances. The basis of an epidemic is not pestilence or catarrh: it is Marseilles in 1721, or Bicêtre in 1780; it is Rouen in 1769, where 'there occurred, during the summer, an epidemic among the children of the nature of bilious catarrhal and putrid fevers complicated by miliaria, and ardent bilious fevers during the autumn. This constitution degenerated into putrid biliousness towards the end of that season and during the winter of 1769 and 1770' [4]. The usual pathological forms are mentioned, but as factors in a complex set of intersections in which their role is analogous to that of the symptom in relation to the disease. The essential basis is determined by the time, the place, the 'fresh, sharp, subtle, penetrating' air of Nîmes in winter [5] or the sticky, thick, putrid air of Paris during a long, heavy summer [6].

The regularity of symptoms does not allow the wisdom of a natural order to show through as in filigree; it treats only the constancy of causes, the obstinacy of a factor whose total, unceasingly repeated pressure determines a preferential form of disease. It may be a cause that survives in time—being responsible, for example, for plica in Poland and scrofula in Spain—in which cases the term endemic will be more readily used; or it may be causes that 'suddenly attack a large number of people in one place, without distinction of age, sex, or temperament. They appear to proceed from a single cause, but as these diseases reign only for a limited period, this cause may be regarded as purely accidental' [7]: this is so in the case of smallpox, malign fever, or dysentery, which are epidemics in the true sense. It is hardly surprising that despite the great diversity, in disposition and age, of the people affected, the disease shows the same symptoms in all: this is because dryness or humidity, heat or cold, when prolonged, ensure the domination of one of our constitutive principles: alkalis, salts, phlogiston; 'We are then exposed to the accidents occasioned by this principle, and these accidents must be the same for different subjects' [8].

The analysis of an epidemic does not involve the recognition of the general form of the disease, by placing it in the abstract space of nosology, but the rediscovery, beneath the general signs, of the particular process, which varies according to circumstances from one epidemic to another, and which weaves from the cause to the morbid form a web common to all the sick, but peculiar to this moment in time and this place in space; in Paris, in 1785, there was an epidemic of quartan fever and putrid synochus, but the essence of the epidemic was that 'the bile had dried up in its passages and turned into melancholy, the blood had become impoverished, thickened, and sticky as it were, the organs of the lower part of the abdomen had swollen and become the causes or centres of obstruction' [9], or a sort of over-all singularity, an individual with many similar heads, whose features are manifested only once in time and space. The specific disease is always more or less repeated, the epidemic is never quite repeated.

In this perceptual structure, the problem of contagion is of little importance. Transmission from one individual to another is never the essence of an epidemic; it may, in the form of 'miasma' or 'leaven', which can be communicated through water, food, contact, the wind, or confined air, constitute one of the causes of the epidemic, either direct or primary (when it is the sole, operant cause), or secondary (when, in a town or hospital, the miasma is the product of an epidemic disease caused by some other factor). But contagion is only one modality of the brute fact of the epidemic. It was readily admitted that malign diseases, like plague, had a transmittable cause; it was more difficult to recognize the same fact in the case of the simple, epidemic diseases (whooping cough, measles, scarlet fever, bilious diarrhoea, intermittent fever) [10].

Whether contagious or not, an epidemic has a sort of historical individuality, hence the need to employ a complex method of observation when dealing with it. Being a collective phenomenon, it requires a multiple gaze; a unique process, it must be described in terms of its special, accidental, unexpected qualities. The event must be described in detail, but it must also be described in accordance with the coherence implied by multi-perception: being an imprecise form of knowledge, insecurely based while ever partial, incapable of acceding of itself to the essential or fundamental, it finds its own range only in the cross-checking of viewpoints, in repeated, corrected information, which finally circumscribes, where gazes meet, the individual, unique nucleus of these collective phenomena. At the end of the eighteenth century, this form of experience was being institutionalized. In each subdelegation a physician and several surgeons were appointed by the *Intendant* (provincial administrator) to study those epidemics that might break out in their canton; they were in constant correspondence with the chief physician of the *généralité* (treasury subdivision of old France) concerning 'both the reigning disease and the medicinal topography of their canton', and when four or five people succumbed to the same disease, the syndic had to notify the subdelegate, who sent the physician to prescribe the treatment to be administered daily by the surgeons; in more serious cases, the physician of the *généralité* visited the scene of the outbreak himself [11].

But this experience could achieve full significance only if it was supplemented by constant, constricting intervention. A medicine of epidemics could exist only if supplemented by a police: to supervise the location of mines and cemeteries, to get as many corpses as possible cremated instead of buried, to control the sale of bread, wine, and meat [12], to supervise the running of abattoirs and dye works, and to prohibit unhealthy housing; after a detailed study of the whole country, a set of health regulations would have to be drawn up that would be read 'at service or mass, every Sunday and holy day', and which would explain how one should feed and dress oneself, how to avoid illness, and how to prevent or cure prevailing diseases: 'These precepts would become like prayers that even the most ignorant, even children, would learn to recite' [13]. Lastly, a body of health inspectors would have to be set up that could be 'sent out to the provinces, placing each one in charge of a particular department'; there he would collect information about the various domains related to medicine, as well as about physics, chemistry, natural history, topography, and astronomy, would prescribe the measures to be taken, and would supervise the work of the doctor. 'It is to be hoped that the state would provide for these physicians and spare them the expense that an inclination to make useful discoveries entails' [14].

A medicine of epidemics is opposed at every point to a medicine of classes, just as the collective perception of a phenomenon that is widespread but unique and unrepeatable may be opposed to the individual perception of the identity of an essence as constantly revealed in the multiplicity of phenomena. The analysis of a series in the one case, the decipherment of a type in the other; the integration of time in the case of epidemics, the determination of hierarchical place in the case of the species; the attribution of a causality —the search for an essential coherence, the subtle perception of a complex historical and geographical space—the demarcation of a homogeneous surface in which analogies can be read. And yet, in the final analysis, when it is a question of these tertiary figures that must distribute the disease, medical experience and the doctor's supervision of social structures, the pathology of epidemics and that of the species are confronted by the same requirements: the definition of a political status for medicine and the constitution, at state level, of a medical consciousness whose constant task would be to provide information, supervision, and constraint, all of which 'relate as much to the police as to the field of medicine proper' [15].

This was the origin of the Société Royale de Médecine and its insuperable conflict with the Faculté (the university authorities). In 1776, the government decided to set up at Versailles a society for the study of the epidemic and epizootic phenomena that had increased considerably in recent years. The precise occasion was a disease affecting livestock that had broken out in southeastern France, and which had forced the Contrôleur Général des Finances to order the killing off of all suspect animals; this led to a fairly serious disruption of the regional economy. The decree of 29 April 1776 declares in its preamble that epidemics are deadly and destructive at the outset only because their character, being little known, leaves the doctor in uncertainty as to the choice of treatment that should be applied; and this uncertainty arises because so little has been done to study the different treatments used, or to describe the symptoms of the different epidemics and the curative methods that have been most successful.

The commission was to have a three-fold role: investigation, by keeping itself informed of the various epidemic movements; elaboration, by comparing facts, recording the treatments used, and organizing experiments; and supervision and prescription, by informing the medical practitioners of the methods that seem to be most suitable to a given situation. It was to be made up of eight doctors: a *directeur*, entrusted with 'the correspondence concerning epidemic and epizootic diseases' (de Lasson), a *commissaire général*, who would co-ordinate the work of the provincial doctors (Vicq d'Azyr), and six doctors of the Faculté, who would devote themselves to work on these same subjects. The Contrôleur des Finances could send them out to the provinces to make inquiries and ask them for reports. Lastly, Vicq d'Azyr was to give a course in human and comparative anatomy to the other members of the commission, the doctors of the Faculté, and 'those students who showed themselves to be worthy of it' [16]. Thus a double check was set up: that of the political authorities over the practice of medicine and that of a privileged medical body over the practitioners as a whole.

The conflict with the Faculté broke out at once. In contemporary eyes, it was a collision of two institutions, one modern and politically supported, the other archaic and inward-looking. A partisan of the Faculté described their opposition thus:

The one ancient, respectable for all manner of reasons and principally in the eyes of the members of the society most of whom have been trained by it; the other, a modern institution whose members have preferred to associate with ministers of the Crown rather than with their own institutions, who have deserted the Assemblies of the Faculté to which the public good and their oaths should have kept them attached for a career of intrigue [17].

For three months, the Faculté 'went on strike' in protest: it refused to exercise its functions, and its members refused to consult with the members of the society. But the outcome was determined in advance because the Conseil supported the new committee. By 1778, the letters patent confirming its transformation into the Société Royale de Médecine had been registered, and the Faculté had been forbidden 'to employ any kind of defence in this affair'. The Société received an income of 40,000 francs raised from mineral waters, while the Faculté received hardly 2,000 francs [18]. But, above all, its role was constantly being enlarged: as a control body for epidemics, it gradually became a point for the centralization of knowledge, an authority for the registration and judgement of all medical activity. At the beginning of the Revolution, the Finance Committee of the National Assembly was to justify its status thus: 'The object of this society is to link French medicine with foreign medicine by means of a useful correspondence; to gather together isolated observations, to preserve them and to compare them; and, above all, to research into the causes of common diseases, to forecast their occurrence, and to discover the most effective remedies for them' [19]. The Société no longer consisted solely of doctors who devoted themselves to the study of collective pathological phenomena; it had become the official organ of a *collective consciousness* of pathological phenomena, a consciousness that operated at both the level of experience and the level of knowledge, in the international as well as the national space.

Political events had a certain novelty value here, as far as basic structures were concerned. A new type of experience was created whose general lines, formed around the years 1775–1780, were to extend far in time and bring with them, during the Revolution and right up to the Consulate, many projects of reform. No doubt very few of these plans were ever implemented. And yet the form of medical perception that they involve is one of the constituent elements of clinical experience.

There was a new style of totalization. The treatises of the eighteenth century, Institutions, Aphorisms, Nosologies, enclosed medical knowledge within a defined space: the table drawn up may not have been complete in every detail, and may have contained gaps here and there owing to ignorance, but in its general form it was exhaustive and closed. It was now replaced by open, infinitely extendable tables. Hautesierck had already provided an example of such a table, when, at Choiseul's request, he proposed a plan of collective work for military physicians and surgeons, comprising four parallel, unlimited series: the study of topographies (location, terrain, water, air, society, the temperaments of the inhabitants), meteorological observations (pressure, temperature, winds), an analysis of epidemics and common diseases, and a description of extraordinary cases [20]. The theme of the encyclopaedia is replaced by that of constant, constantly revised information, where it is a question, rather, of totalizing events and their determination than of enclosing knowledge in a systematic form: 'It is so true that there exists a chain linking, throughout the universe, on earth and in man, all beings, all bodies, all affections; a chain whose subtlety eludes the superficial gaze of the meticulous experimenter and the writer of cold dissertations, but is revealed to the truly observant genius' [21]. At the beginning of the Revolution, Cantin proposed that this work of information should be undertaken in each department by a commission elected from among the doctors [22]; Mathieu Géraud demanded the creation in every large town of a 'government health centre' and in Paris of a 'health court', sitting beside the National Assembly, centralizing information, conveying it from one part of the country to another, discussing questions that still remain obscure, and indicating what research needs to be carried out [23].

What now constituted the unity of the medical gaze was not the circle of knowledge in which it was achieved but that open, infinite, moving totality, ceaselessly displaced and enriched by time, whose course it began but would never be able to stop—by this time a clinical recording of the infinite, variable series of events. But its support was not the perception of the patient in his singularity, but a collective consciousness, with all the information that intersects in it, growing in a complex, ever-proliferating way until it finally achieves the dimensions of a history, a geography, a state.

In the eighteenth century, the fundamental act of medical knowledge was the drawing up of a 'map' (repérage): a symptom was situated within a disease, a disease in a specific ensemble, and this ensemble in a general plan of the pathological world. In the experience that was being constituted towards the end of the century, it was a question of 'carving up' the field by means of the interplay of series, which, in intersecting one another, made it possible to reconstitute the chain referred to by Menuret. Each day Razoux made meteorological and climatic observations, which he then compared with a nosological analysis of patients under observation and with the evolution, crises, and outcome of the diseases [24]. A system of coincidences then appeared that indicated a causal connexion and also suggested kinships or new links between diseases.

'If anything is able to improve our art,' Sauvages himself wrote to Razoux, 'it is work of this kind carried out over a period of fifty years, by a team of thirty doctors as meticulous and industrious as yourself. . . . I will do all in my power to have one of our doctors carry out the same observations in our Hôtel-Dieu' [25]. What defines the act of medical knowledge in its concrete form is not, therefore, the encounter between doctor and patient, nor is it the confrontation of a body of knowledge and a perception; it is the systematic intersection of two series of information, each homogeneous but alien to each other—two series that embrace an infinite set of separate events, but whose intersection reveals, in its isolable dependence, the *individual fact*. A sagittal figure of knowledge.

In this movement, medical consciousness is duplicated: it lives at an immediate level, in the order of 'savage' observations; but it is taken up again at a higher level, where it recognizes the constitutions, confronts them, and, turning back upon the spontaneous forms, dogmatically pronounces its judgement and its knowledge. It becomes centralized in structure. At the institutional level this is apparent in the Société Royale de Médecine. And at the beginning of the Revolution there were innumerable projects that schematized this dual and necessary authority (*instance*) of medical knowledge, with its ceaseless movement between these two levels, at the same time maintaining and traversing the distance between them. Mathieu Géraud proposed the setting up of a Health Court (*Tribunal de Salubrité*) where a prosecutor would denounce 'any person who, without having given proof of his ability, exercises upon another, or upon an animal that does not belong to him, anything pertaining to the direct or indirect application of the art of health' [26]; the decisions of this court concerning professional abuses, inadequacies, and imperfections should constitute the jurisprudence of the medical state. In addition to a Judiciary, there should be an Executive that would exercise a policing function over all aspects of health (*la haute et grande police sur toutes les branches de la salubrité*). It would prescribe what books were to be read and what new works were to be written; it would indicate, on the basis of the information received, what treatment was to be administered for prevalent diseases; it would publish whatever was required by an enlightened medical practice, whether the results of inquiries carried out under its own supervision or foreign works. Following an autonomous movement, the medical gaze circulates within an enclosed space in which it is controlled only by itself; in sovereign fashion, it distributes to daily experience the knowledge that it has borrowed from afar and of which it has made itself both the point of concentration and the centre of diffusion.

In that experience, medical space can coincide with social space, or, rather, traverse it and wholly penetrate it. One began to conceive of a generalized presence of doctors whose intersecting gazes form a network and exercise at every point in space, and at every moment in time, a constant, mobile, differentiated supervision. The problem of the settling of doctors in the countryside was raised [27]; there were requests for a statistical supervision of health based on the registration of births and deaths (which would have to mention the disease from which the individual suffered, his mode of life, and the cause of his death, thus constituting a pathological record); there were demands that the reasons for exemption from military service on medical grounds should be given in detail by the recruiting board; in fact, that a medical topography of each department should be drawn up, 'with detailed observations concerning the region, housing, people, principal interests, dress, atmospheric constitution, produce of the ground, time of their perfect maturity and their harvesting, and physical and moral education of the inhabitants of the area' [28]. And since the question of the settling of doctors was not enough, the consciousness of each individual must be alerted; every citizen must be informed of what medical knowledge is necessary and possible. And each practitioner must supplement his supervisory activity with teaching, for the best way of avoiding the propagation of disease is to spread medical knowledge [29]. The locus in which knowledge is formed is no longer the pathological garden where God distributed the species, but a generalized medical consciousness, diffused in space and time, open and mobile, linked to each individual existence, as well as to the collective life of the nation, ever alert to the endless domain in which illness betrays, in its various aspects, its great, solid form.

The years preceding and immediately following the Revolution saw the birth of two great myths with opposing themes and polarities: the myth of a nationalized medical profession, organized like the clergy, and invested, at the level of man's bodily health, with powers similar to those exercised by the clergy over men's souls; and the myth of a total disappearance of disease in an untroubled, dispassionate society restored to its original state of health. But we must not be misled by the manifest contradiction of the two themes: each of these oneiric figures expresses, as if in black and white, the same picture of medical experience. The two dreams are isomorphic: the first expressing in a very positive way the strict, militant, dogmatic medicalization of society, by way of a quasi-religious conversion, and the establishment of a therapeutic clergy; the second expressing the same medicalization, but in a triumphant, negative way, that is to say, the volatilization of disease in a corrected, organized, and ceaselessly supervised environment, in which medicine itself would finally disappear, together with its object and its *raison d'être*.

Sabarot de l'Avernière, a prolific author of projects in the early years of the Revolution, saw priests and doctors as the natural heirs of the Church's two most visible missions—the consolation of souls and the alleviation of pain. So the wealth of the Church, which has been diverted from its original use by the higher clergy, must be confiscated and returned to the nation, which alone knows its own spiritual and material needs. The revenues would be divided equally between the parish clergy and the doctors. Are not doctors the priests of the body? 'The soul cannot be considered separately from animate bodies, and if the ministers of the Altars are venerated, and receive from the state a reasonable living, those who tend your health should also receive a salary sufficient to feed themselves and to succour you. They are the tutelary genii of the integrity of your faculties and sensations' [30]. The doctor would no longer have to demand a fee from his patient; the treatment of the sick would be free and obligatory—a service that the nation would provide as one of its sacred tasks; the doctor would be no more than the instrument of that service [31]. At the end of his studies, the new doctor would occupy not the post of his choice, but the one that was assigned to him according to the needs and vacancies, throughout the country; when he had gained in experience, he could apply for a more responsible, better-paid job. He would have to give an account to his superiors of his activities and would be held responsible for his mistakes. Having become a public, disinterested, supervised activity, medicine could improve indefinitely; in the alleviation of physical misery, it would be close to the old spiritual vocation of the Church, of which it would be a sort of lay carbon copy. To the army of priests watching over the salvation of souls would correspond that

of the doctors who concern themselves with the health of bodies.
The other myth proceeds from a historical reflexion carried to
its conclusion. Linked as they are with the conditions of existence
and with the way of life of individuals, diseases vary from one
period and one place to another. In the Middle Ages, at a time of
war and famine, the sick were subject to fear and exhaustion (apoplexy, hectic fever); but in the sixteenth and seventeenth centuries,
a period of relaxation of the feeling for one's country and of the
obligations that such a feeling involves, egotism returned, and lust
and gluttony became more widespread (venereal diseases, congestion of the viscera and of the blood); in the eighteenth century, the
search for pleasure was carried over into the imagination: one went
to the theatre, read novels, and grew excited in vain conversations;
one stayed up at night and slept during the day (hysteria, hypo-
chondria, nervous diseases) [32]. A nation that lived without war,
without violent passions, without idleness would know none of
these ills, nor, above all, would a nation that did not know the
tyranny of wealth over poverty, nor given to abuses. The rich?
'Living in the midst of ease, surrounded by the pleasures of life,
their irascible pride, their bitter spleen, their abuses, and the excesses
to which their contempt of all principles leads them makes them
prey to infirmities of every kind; soon . . . their faces are furrowed,
their hair turns white, and diseases harvest them before their time)
[33]. Meanwhile, the poor, subjected to the despotism of the rich
and of their kings, know only taxes that reduce them to penury,
scarcity that benefits only the profiteers, and unhealthy housing
that forces them 'either to refrain from raising families or to pro-
create weak, miserable creatures' [34].

The first task of the doctor is therefore political: the struggle
against disease must begin with a war against bad government. Man
will be totally and definitively cured only if he is first liberated:
'Who, then, should denounce tyrants to mankind if not the doctors,
who make man their sole study, and who, each day, in the homes
of poor and rich, among ordinary citizens and among the highest in
the land, in cottage and mansion, contemplate the human miseries
that have no other origin but tyranny and slavery?' [35]. If medicine could be politically more effective, it would no longer be indispensable medically. And in a society that was free at last, in which
inequalities were reduced, and in which concord reigned, the doctor would have no more than a temporary role: that of giving

legislator and citizen advice as to the regulation of his heart and body. There would no longer be any need for academies and hospitals:

> By training citizens in frugality by means of simple dietary laws, by showing young people above all the pleasures that may be derived from even a hard life, by making them appreciate the strictest discipline in the army and navy, how many ills would be prevented, how much expense avoided, and what new abilities would reveal themselves . . . for the greatest, most difficult enterprises.

And gradually, in this young city entirely dedicated to the happiness of possessing health, the face of the doctor would fade, leaving a faint trace in men's memories of a time of kings and wealth, in which they were impoverished, sick slaves.

All this was so much day-dreaming; the dream of a festive city, inhabited by an open-air mankind, in which youth would be naked and age know no winter, the familiar symbol of ancient arcadias, to which has been added the more recent theme of a nature encompassing the earliest forms of truth—all these values were soon to fade [36].

And yet they played an important role: by linking medicine with the destinies of states, they revealed in it a positive significance. Instead of remaining what it was, 'the dry, sorry analysis of millions of infirmities', the dubious negation of the negative, it was given the splendid task of establishing in men's lives the positive role of health, virtue, and happiness; it fell to medicine to punctuate work with festivals, to exalt calm emotions, to watch over what was read in books and seen in theatres, to see that marriages were made not out of self-interest or because of a passing infatuation, but were based on the only lasting condition of happiness, namely, their benefit to the state [37].

Medicine must no longer be confined to a body of techniques for curing ills and of the knowledge that they require; it will also embrace a knowledge of *healthy man*, that is, a study of *non-sick man* and a definition of the *model man*. In the ordering of human existence it assumes a normative posture, which authorizes it not only to distribute advice as to healthy life, but also to dictate the standards for physical and moral relations of the individual and of the society in which he lives. It takes its place in that borderline, but for modern man paramount, area where a certain organic, unruffled, sensory happiness communicates by right with the order of a nation, the vigour of its armies, the fertility of its people, and the patient advance of its labours. The dreamer Lanthenas gave medicine a definition that was brief but heavy with history: 'At last, medicine will be what it must be, the knowledge of natural and social man' [38].

It is important to determine how and in what manner the various forms of medical knowledge pertained to the positive notions of 'health' and 'normality'. Generally speaking, it might be said that up to the end of the eighteenth century medicine related much more to health than to normality; it did not begin by analysing a 'regular' functioning of the organism and go on to seek where it had deviated, what it was disturbed by, and how it could be brought back into normal working order; it referred, rather, to qualities of vigour, suppleness, and fluidity, which were lost in illness and which it was the task of medicine to restore. To this extent, medical practice could accord an important place to regimen and diet, in short, to a whole rule of life and nutrition that the subject imposed upon himself. This privileged relation between medicine and health involved the possibility of being one's own physician. Nineteenth-century medicine, on the other hand, was regulated more in accordance with normality than with health; it formed its concepts and prescribed its interventions in relation to a standard of functioning and organic structure, and physiological knowledge—once marginal and purely theoretical knowledge for the doctor—was to become established (Claude Bernard bears witness to this) at the very centre of all medical reflexion. Furthermore, the prestige of the sciences of life in the nineteenth century, their role as model, especially in the human sciences, is linked originally not with the comprehensive, transferable character of biological concepts, but, rather, with the fact that these concepts were arranged in a space whose profound structure responded to the healthy/morbid opposition. When one spoke of the life of groups and societies, of the life of the race, or even of the 'psychological life', one did not think first of the internal structure of *the organized being*, but of *the medical bipolarity of the normal and the pathological*. Consciousness lives because it can be altered, maimed, diverted from its course, paralysed; societies live because there are sick, declining societies and healthy, expanding ones; the race is a living being that one can see degenerating; and civilizations, whose deaths have so often been remarked on, are also, therefore, living beings. If the science of man appeared as an extension of the science of life, it is because it was *medically*, as well as *biologically*, based: by transference, importation, and, often, metaphor, the science of man no doubt used concepts formed by biologists; but the very subjects that it devoted itself to (man, his behaviour, his individual and social realizations) therefore opened up a field that was divided up according to the principles of the normal and the pathological. Hence the unique character of the science of man, which cannot be detached from the negative aspects in which it first appeared, but which is also linked with the positive role that it implicitly occupies as norm.

NOTES

[1] Th. Sydenham, 'Observationes medicae', *Opera medica* (Geneva, 1736, I, p. 32).
[2] *Ibid.*, p. 27.
[3] Le Brun, *Traité historique sur les maladies épidémiques* (Paris, 1776, p. 1).
[4] Lepecq de la Clôture, *Collection d'observations sur les maladies et constitutions épidémiques* (Rouen, 1778, p. xiv).
[5] Razoux, *Tableau nosologique et météorologique* (Basel, 1787, p. 22).
[6] Menuret, *Essai sur l'histoire médico-topographique de Paris* (Paris, 1788, p. 139).
[7] Banan and Turben, *Mémoires sur les épidémies du Languedoc* (Paris, 1786, p. 3).
[8] Le Brun, *op. cit.*, p. 66, n. 1.
[9] Menuret, *op. cit.*, p. 139.
[10] Le Brun, *op. cit.*, pp. 2–3.
[11] Anon., *Description des épidémies qui ont régné depuis quelques années sur la généralité de Paris* (Paris, 1783, pp. 35–7).
[12] Le Brun, *op. cit.*, pp. 127–32.
[13] Anon., *op. cit.*, pp. 14–17.
[14] Le Brun, *op. cit.*, p. 124.
[15] *Ibid.*, p. 126.
[16] Cf. *Précis historique de l'établissement de la Société royale de Médecine* (Undated. The anonymous author is Boussu).
[17] Retz, *Exposé succint à l'Assemblée Nationale* (Paris, 1791, pp. 5–6).
[18] Cf. Vacher de la Feuterie, *Motif de la réclamation de la Faculté de Médecine de Paris contre l'établissement de la Société royale de Médecine* (Place and date of publication unknown).
[19] Quoted in Retz, *op cit.*
[20] Hautesierck, *Recueil d'observations de médecine des hôpitaux militaires* (Paris, 1766, vol. I, pp. xxiv–xxvii).

[21] Menuret, *op cit.*, p. 139.

[22] Cantin, *Projet de réforme adressé à l'Assemblée Nationale* (Paris, 1790).

[23] Mathieu Géraud, *Projet de décret à rendre sur l'organisation civile des médecins* (Paris, 1791, nos. 78–9).

[24] Razoux, *op. cit.*

[25] Quoted in *ibid.*, p. 14.

[26] Mathieu Géraud, *op. cit.*, p. 65.

[27] Cf. N.-L. Lespagnol, *Projet d'établir trois médecins par district pour le soulagement des gens de la campagne* (Charleville, 1790). Royer, *Bienfaisance médicale et projet financier* (Provins, Year IX).

[28] J.-B. Demangeon, *Des moyens de perfectionner la médecine* (Paris, Year VII, pp. 5–9); cf. Audin Rouvière, *Essai sur la topographie physique et médicale de Paris* (Paris, Year II).

[29] Bacher, *De la médecine considérée politiquement* (Paris, Year XI, p. 38).

[30] Sabarot de L'Avernière, *Vue de Législation médicale adressée aux États généraux* (1789, p. 3).

[31] In Menuret, *Essai sur le moyen de former de bons médecins* (Paris, 1791), one finds the idea of financing medicine from church revenues, but only when it is a question of treating the poor.

[32] Maret, *Mémoire où on cherche à déterminer quelle influence les moeurs ont sur la santé* (Amiens, 1771).

[33] Lanthenas, *De l'influence de la liberté sur la santé* (Paris, 1792, p. 8).

[34] *Ibid.*, p. 4.

[35] *Ibid.*, p. 8.

[36] On 2 June 1793, Lanthenas, who was a Girondist, was put on the proscribed list, then crossed off, Marat having described him as 'weakheaded'. Cf. Mathiez, *La Révolution française* (Paris, 1945, vol II, p. 221).

[37] Cf. Ganne, *De l'homme physique et moral, ou recherches sur les moyens de rendre l'homme plus sage* (Strasbourg, 1791).

[38] Lanthenas, *op. cit.*, p. 18.

3 · The Free Field

The contrast between a medicine of pathological spaces and a medicine of the social space was concealed from contemporaries by the visible prestige accorded to a consequence common to both: the removal from the field of all medical institutions that proved unyielding towards the new requirements of the gaze. In fact, an entirely free field of medical experiment had to be constituted, so that the natural needs of the species might emerge unblurred and without trace; it also had to be sufficiently present in its totality and concentrated in its content to allow the formation of an accurate, exhaustive, permanent corpus of knowledge about the health of a population. This medical field, restored to its pristine truth, pervaded wholly by the gaze, without obstacle and without alteration, is strangely similar, in its implicit geometry, to the social space dreamt of by the Revolution, at least in its original conception: a form homogeneous in each of its regions, constituting a set of equivalent items capable of maintaining constant relations with their entirety, a space of free communication in which the relationship of the parts to the whole was always transposable and reversible.

There is, therefore, a spontaneous and deeply rooted convergence between the requirements of *political ideology* and those of *medical technology*. In a concerted effort, doctors and statesmen demand, in a different vocabulary but for essentially identical reasons, the suppression of every obstacle to the constitution of this new space: the hospitals, which alter the specific laws governing

38

disease, and which disturb those no less rigorous laws that define the relations between property and wealth, poverty and work; the association of doctors which prevents the formation of a centralized medical consciousness, and the free play of an experience that is allowed to reach the universal without imposed limitations; and, lastly, the Faculties, which recognize that which is true only in theoretical structures and turn knowledge into a social privilege. Liberty is the vital, unfettered force of truth. It must, therefore, have a world in which the gaze, free of all obstacle, is no longer subjected to the immediate law of truth: the gaze is not faithful to truth, nor subject to it, without asserting, at the same time, a supreme mastery: the gaze that sees is a gaze that dominates; and although it also knows how to subject itself, it dominates its masters:

> Despotism has need of darkness, but liberty, radiant with glory, can only survive when surrounded by all the light that can enlighten men; it is during the sleep of peoples that tyranny can establish itself and become naturalized among them. . . . Make other nations tributaries not of your political authority, nor of your government, but of your talents and your knowledge. . . . There is a dictatorship for peoples whose yoke is not repugnant to those who bend under it, and that is the dictatorship of genius [1].

The ideological theme that guides all structural reforms from 1789 to Thermidor Year II is that of the sovereign liberty of truth: the majestic violence of light, which is in itself supreme, brings to an end the bounded, dark kingdom of privileged knowledge and establishes the unimpeded empire of the gaze.

I. THE INVESTMENT IN HOSPITAL STRUCTURES

The Comité de Mendicité de l'Assemblée Nationale was under the influence of both economists and doctors who believed that the only possible locus for recovering from disease was the natural environment of social life, the family. There the cost of sickness to the nation was reduced to a minimum, and the risk of the disease leading to artificial complications, spreading of its own accord, and assuming, as in hospitals, the aberrant form of a disease of the disease was avoided. In the family, the disease was in a state of 'nature', that is, in accord with its own nature and freely exposed to the regenerative forces of nature. The gaze that is turned upon it by those close to the sick person has the vital force of benevolence and the discretion of hope. In the freely observed disease, there is something that compensates for it:

> Misfortune . . . arouses by its presence beneficent compassion, brings to birth in men's hearts the pressing need to offer comfort and consolation, and the care given to the unfortunate in their own dwellings turns to account that abundant spring of wealth distributed by private benevolence. If the poor man is put into a hospital, he is deprived of all these resources . . . [2].

No doubt there are sick persons who have no family, and others who are so poor that they live 'cooped up in attics'. For these, 'communal houses for the sick' must be set up that would function as family substitutes and spread, in the form of reciprocity, the gaze of compassion; in this way, the poor would find 'in companions of their own kind naturally sympathetic creatures who are at least not entirely strangers to them' [3]. Thus disease would everywhere find its natural, or almost natural, locale, where it would be free to follow its own course and to abolish itself in its truth.

But the ideas of the Comité de Mendicité are also related to the theme of a social, centralized consciousness of disease. A generalized state of health is not to be expected solely from such a freedom. If the family was bound to the unfortunate individual by the *natural* duty of compassion, the nation was bound to him by the *social, collective* duty to provide assistance. Hospital foundations represented an immobilization of wealth, and, by their very inertia, created poverty; these must disappear, but they must be replaced by a national, constantly available fund capable of providing help when and where required. The state must therefore 'divert to its own use' the wealth of the hospitals and then combine it into a 'common fund'. A central body would be set up to administer this fund; it would act as the permanent medico-economic conscience of the nation; it would be the universal perception of every illness and the immediate recognition of all needs. The great *Oeil de la Misère*. It would be given the task of 'distributing sums necessary and completely adequate for the alleviation of the unfortunate'. It would finance the 'communal house' and provide special help to poor families who care for their sick themselves.

The project failed on account of two technical problems. The first, that of the diversion of hospital funds, is political and economic in nature. The second is medical in nature and concerns complex or contagious diseases. The Legislative Assembly went back on the principle of the nationalization of hospital capital; it preferred simply to divert its revenue into an assistance fund. It also decided not to entrust the administration of the fund to a central body, which, it was believed, would be too cumbersome, too distant, and therefore unable to respond to immediate needs. If the consciousness of disease and poverty was to be immediate and effective, it should be a geographically specific consciousness. And in this field, as in so many others, the Legislative Assembly went back on the centralization of the Constituent Assembly and adopted a much looser, Anglicized system: local authorities would make the essential links, keep themselves informed of needs, and distribute the revenues; they would form a multiple network of supervision. Thus the principle of the communalization of assistance was raised —a principle to which the Directoire finally rallied.

But in this dispersed structure, decentralization is associated with two historically important themes—of assistance and of repression. Tenon, in his concern to settle the question of Bicêtre and Salpêtrière [4], wanted the Legislature to create a committee for 'hospitals and houses of arrest' (*maisons d'arrestation*) that would be generally responsible for hospitals, prisons, vagabondage, and epidemics. The Assembly opposed the suggestion on the ground that 'in a sense it debases the lower classes of the people by entrusting the care of the unfortunate and of criminals to the same persons' [5]. The consciousness of disease, and of the assistance that it required among the poor, assumed autonomy; it was now concerned with a very specific type of poverty. Similarly, the doctor began to play a decisive role in the organization of assistance. At the social level at which help was distributed, it was the doctor who discovered where it was needed and judged the nature and degree of the assistance to be given. The decentralization of the means of assistance authorized a medicalization of its distribution. This is reminiscent of an idea made familiar by Cabanis, that of the doctor-magistrate, to whom 'men's lives' would be entrusted by the community instead of 'leaving them to the mercy of mountebanks and gossips'; he would act according to the belief that 'the lives of the rich and powerful are no more precious than those of the poor and weak'; lastly, he would be able to refuse help to 'public malefactors' [6]. In addition to his role as a technician of medicine, he would play an economic role in the distribution of help, and a moral, quasi-judicial role in its attribution; he would become 'the guardian of public morals and public health alike' [7].

In this regional configuration, in which the medical consciousness is made up of discontinuous 'authorities' (*instances*), the hospital must have a place. It is needed for the sick who have no family, but it is also needed in cases of contagion, and for difficult, complex, 'extraordinary' patients with whom medicine in its ordinary, everyday form cannot cope. Again, one can detect the influence of Tenon and Cabanis. The hospital, which, in its general form, was associated only with penury, appears at the local level as an indispensable measure of protection. Protection of the healthy against disease; protection of the sick against the nostrums of the ignorant—'the people must be saved from its own errors' [8]; protection of the sick from one another. What Tenon is proposing is a differentiated hospital space. And differentiated according to two principles: 'training', by which each hospital would devote itself to the care of a particular category of patient or family of diseases; and 'distribution', which, within a single hospital, would determine the order in which 'the different kinds of patient would be arranged with a view to admission' [9]. Thus the family, the natural locus of disease, is duplicated by another space that must reproduce, like a microcosm, the specific configuration of the pathological world. There, beneath the eye of the hospital doctor, diseases would be grouped into orders, genera, and species, in a rationalized domain that would restore the original distribution of essences. Thus conceived, the hospital would make it possible 'to classify patients to such a point that each would find what was suited to his state without aggravating by his proximity the illness of others, and without spreading contagion, either in the hospital or outside it' [10]. In the hospital, disease meets, as it were, the forced residence of its truth.

In the projects of the Comité des Secours, two authorities are juxtaposed: the ordinary, which, because of the distribution of aid, involves a continuous supervision of the social space with a system of highly medicalized regional centres; and the extraordinary, which is made up of discontinuous, exclusively medical spaces, structured according to the model of scientific knowledge. Disease is thus caught in a double system of observation: there is a gaze that does

not distinguish it from, but re-absorbs it into, all the other social
ills to be eliminated; and a gaze that isolates it, with a view to cir-
cumscribing its natural truth.

The Legislative Assembly left to the Convention two problems
that were not resolved: that of the ownership of hospital funds
and the new problem of the staffing of hospitals. On 18 August
1792, the Assembly had declared the dissolution of 'all religious
corporations and secular congregations of men or women, ecclesi-
astical or lay' [11]. But most of the hospitals were run by religious
orders, or, like La Salpêtrière, by lay organizations conceived on a
quasi-monastic model. So the decree added: 'Nevertheless, in hospitals and houses of charity, the same persons will continue as before
to serve the poor and care for the sick in an individual capacity,
under the supervision of the municipal and administrative bodies,
until their definitive organization is presented to the National Assembly by the Comité des Secours.' In fact, right up to the fall of
Robespierre (9 Thermidor), the Convention was to consider the
problem of assistance and the hospitals, above all, in terms of abolition. The immediate abolition of state help demanded by the
Girondists, who feared the political adhesion of the poorest classes
to the Communes, if the latter were given the task of distributing
assistance. For Roland, the system of 'handouts' was 'the most
dangerous one': no doubt beneficence can and must be carried out
by 'private subscription, but the government must not interfere;
it would be misled and would give little or no help' [12]. The
abolition of the hospitals was demanded by the Mountain, the ex-
tremist party, who regarded them as an institutionalization of poverty and who believed that one of the tasks of the Revolution must
be to make them unnecessary. Speaking of a hospital devoted 'to
suffering humanity', Lebon asked: 'Must any section of mankind be
sick and needy? . . . Therefore let notices be placed over the gates
of these asylums announcing their coming disappearance. For if
when the Revolution is complete we still have such unfortunates
amongst us, our revolutionary work will have been in vain' [13].
And Barère, in the debate of the Law of 22 Floréal (April–May),
Year II, was to launch the famous cry: 'No more alms, no more
hospitals!'

With the victory of the Mountain, the idea of an organization
of public assistance by the state and of a complementary abolition
of the hospitals, over a fairly long period of time, was accepted.

The constitution of Year II proclaims in its Declaration of Rights that 'public assistance is a sacred debt'; the Law of 22 Floréal ordered the drawing up of 'a great book of national beneficence' and the organization of a system of help throughout the country-side. Provision was made for 'houses of health' only for 'the sick who have no home or who cannot receive help there' [14]. The nationalization of hospital funds, which had been accepted in prin-ciple since 19 March 1793, but the application of which was to be postponed until after 'a complete, definitive organization in several areas of public help', was put into immediate execution with the Law of 23 Messidor (June–July), Year II. The hospital funds would be regarded as national property, and assistance would be the responsibility of the Treasury. Cantonal agencies would be entrusted with the task of distributing the help needed by each household. Thus, in legislation if not in reality, the great dream of a total dehospitalization of disease and poverty began to be brought about. Poverty is an economic fact for which assistance must be given while it exists; disease is an individual accident that the family must respond to by ensuring that the victim has the necessary care. The hospital is an anachronistic solution that does not respond to the real needs of the poor and that stigmatizes the sick in a state of penury. There must be an ideal state in which the human being would no longer know exhaustion from hard labour or the hospital that leads to death. 'A man is made neither for a trade, nor for a hospital, nor for a poorhouse: such a prospect is too terrible' [15].

II. THE LAW OF MEDICAL PRACTICE AND TEACHING

The decrees of Marly, issued in March 1707, regulated the practice of medicine and the training of doctors for the rest of the eight-eenth century. It was then a matter of struggling against charlatans, quacks, and 'unqualified and incapable persons practising medicine'; similarly, there had been a need to reorganize the medical faculties, which for many years had fallen into the most 'extreme slackness'. It was laid down that henceforth medicine would be taught in all the universities of the kingdom that had, or had had, a faculty; that the chairs, instead of remaining vacant for an indefinite period, would be made available as soon as they became free; that the students would receive their degree only after three years of study, duly verified by matriculation every four months; that every year

they would have to pass an examination before receiving the title
of *bachelier, licencié*, or *docteur*; that they would follow compul-
sory courses in anatomy, in chemical and Galenic pharmacy, and
in demonstrations of plants [16]. In these conditions, Article 26
of the decree enunciated the principle that 'no person may practise
medicine, or prescribe any remedy, even without payment, if he
has not obtained the degree of *licencié*'; and the text added—and
this was the fundamental issue and aim achieved by the Faculties
of Medicine at the cost of their reorganization—'And all religious,
mendicant or non-mendicant, shall be and remain included in the
prohibition laid down in the preceding article' [17]. By the end of
the century, the critics were unanimous on at least four points:
charlatans continued to flourish; the canonical teaching provided
by the Faculties no longer satisfied either the needs of medical practice or new discoveries (only theory was taught; neither mathematics nor physics was considered); there were too many schools
of medicine for teaching to be carried out in a satisfactory manner;
peculation was rife (the chairs were obtained like any other post:
the professors charged for their lectures, the students bought their
examinations and got needy doctors to write their theses for them),
which made medical studies extremely costly—a situation made
worse by the fact that, when qualified, the new doctor still had to
gain practical experience by accompanying some well-known practitioner on his visits, for which privilege he again had to pay [18].
The Revolution was faced, therefore, with two demands: a stricter
limitation of the right to practise and a stricter organization of the
university cursus. But both went against the whole movement of
reforms that culminated in the abolition of guilds and the master/
apprentice system and in the closing of the universities.

There was thus a certain amount of tension between the requirements of a reorganization of knowledge, those of the abolition
of privileges and those of an effective supervision of the nation's
health. How can the free gaze that medicine, and, through it, the
government, must turn upon the citizens be equipped and competent without being embroiled in the esotericism of knowledge
and the rigidity of social privilege?

First problem: Can medicine be a free profession that is protected by no corporative law, no prohibition of practice, no privilege
of qualification? Can the medical consciousness of a nation be as
spontaneous as its civic or moral consciousness? Doctors defend

their corporate rights on the ground that they should be understood
not in the sense of privilege but of collaboration. The medical body
is to be distinguished from political bodies in that it does not seek
to limit the liberty of others or to impose laws and obligations upon
the citizens; its imperative applies only to itself; its 'jurisdiction is
concentrated within itself' [19]; but it is also to be distinguished
from other professional bodies because it is intended not to preserve
rights and obscure traditions but to confront and to communicate
knowledge: without an established organ, enlightenment would be
extinguished at birth and individual experience lost for all. In forming themselves into a body, doctors make the following implicit
oath: 'We wish to enlighten our minds by fortifying ourselves with
our collective knowledge; the weakness of some of our number is
corrected by the superiority of others; by coming together under
a common administration we will continue to arouse competition
among ourselves' [20]. The medical body criticizes itself to a
greater extent than it protects itself, and, by virtue of this fact, it is
indispensable in protecting the people from its own illusions and
from the mystifications of charlatans [21]. 'If physicians and surgeons form a necessary body in society, their important functions
require on the part of the legislative authority special consideration
in the prevention of abuses' [22]. A free state that wishes to maintain its citizens free from error and from the ills that it entails cannot authorize the free practice of medicine.

In fact, no one, not even the most liberal of the Girondists,
dreamt of freeing medical practice entirely and opening it up to a
free regime of uncontrolled competition. While demanding the
abolition of all constituted medical bodies, even Mathieu Géraud
wished to set up in each department a court that would try 'any
private person dabbling in medicine without having given proof of
his skill' [23]. But the problem of the practice of medicine was
linked to three other problems: the general abolition of guilds, the
disappearance of the society of medicine, and, above all, the closing
of the universities.

Up to Thermidor, there were innumerable projects for the reorganization of the Schools of Medicine. They fall into two groups,
the first presupposing the survival of university structures and the
second taking into account the decrees of 17 August 1792. Among
the 'reformists' one constantly encounters the idea that local interests must be abolished, together with the smaller, moribund

Faculties, in which an inadequate number of professors, all incompetent, distribute or sell degrees and other qualifications. A small number of Faculties would offer chairs throughout the country that would be filled by the best candidates; they would train doctors whose quality would be undisputed; the double-check of the state and public opinion would thus favour the development of a body of medical knowledge and a medical consciousness that would at last be adequate to the nation's needs. Thiery thought that four Faculties would be enough; Gallot preferred two, with a number of special schools for a less-advanced course of training [24]. Moreover, the duration of studies would have to be longer: seven years according to Gallot, ten according to Cantin; this was because it was now intended to include in the curriculum mathematics, geometry, physics, and chemistry [25], all of which had an organic connexion with medical science. But, above all, there had to be practical training. Thiery wanted a Royal Institute, which would provide the pick of the young doctors with a more advanced, essentially practical training; a sort of residential school would be set up in the Jardin du Roi that would operate in close co-operation with a hospital (La Salpêtrière, which was nearby, would serve the purpose); there the professors would teach as they visited the patients; the Faculté would merely appoint a doctor-regent for the public examinations of the Institute. Cantin suggested that once the rudiments had been taught, the candidate doctors would be sent either to a hospital or to the countryside, where they would attain practical experience as assistants to already qualified doctors; for very often what is needed is an extra pair of hands, and patients rarely need highly qualified doctors. By making a kind of medical tour of France, the future doctors would acquire the most varied experience, learn to recognize the diseases peculiar to each climate, and learn what methods were most successful in the treatment of each illness.

A practical training curiously independent of the theoretical teaching provided by the university was proposed. Whereas, as we shall see later, medicine already possessed concepts that enabled it to define the unity of clinical teaching, the theoreticians failed to propose an institutional version of it: practical training was not simply the application of abstract knowledge (if so, it would be enough to entrust this practical teaching to the professors in the schools); nor could it be the key to this knowledge (which could be acquired only when this knowledge had been mastered), because, in fact, this practical teaching still concealed the technological structure of a medicine of the social group, whereas the university training was inseparable from a medicine that was so closely related to the theory of species.

In a rather paradoxical way, this acquisition of practical training, which is dominated by the theme of social usefulness, was left almost entirely to private initiative, with the state controlling little more than the theoretical teaching. Cabanis wanted every hospital doctor to be allowed to 'form a school according to whatever plan he considered most suitable'. He and he alone would decide the duration of each student's studies: for some, two years would be enough; other, less gifted students would require four years. As the result of individual initiative, these lessons would have to be paid for by the students, and the professors themselves would determine the fees; these might be very high in the case of a famous professor whose teaching was much in demand, but this would be no bad thing: 'a spirit of noble emulation, sustained by all manner of motives, cannot but be to the advantage of patients, students, and science' [26].

This reformist thinking has a curious and complex structure. Assistance was to be left to individual initiative, and the hospital establishments were to be maintained for a more complex, almost privileged medicine; by a kind of exchange of places, the position of teaching was inverted. It followed an obligatory, public course to the university, to become, at the hospital stage, private, competitive, and fee-paying; this was because at this level the technological structures of knowledge and that of perception were not yet capable of being superimposed: the way in which one directed one's gaze and the way in which it was trained did not overlap. The field of practical medicine was divided between a free, endlessly open domain—that of home practice—and a closed space, confined to the truths of the species that it revealed; the field of apprenticeship was divided between an enclosed domain of essential truths and a free domain in which truth speaks of itself. And the hospital played this dual role: for the doctor's gaze it was the locus of systematic truths; for the knowledge formulated by the teacher it was the locus of free experiment.

In August 1791, the universities were closed down; in September the Legislative Assembly was dissolved. The ambiguity of these complex structures was about to end. The Girondists demanded total freedom, and they were supported by all those who had benefitted from the old state of affairs and who, in the absence of any organization, thought that they might get back, if not their privileges, at least their influence. Catholics like Durand Maillane, former Oratorian fathers like Daunou or Sieyès, moderates like Fourcroy were all advocates of extreme liberalism in teaching arts and sciences. For them, Condorcet's project threatened to reconstitute a 'formidable corporation' [27]; there would be a rebirth of what had only recently been abolished, 'the Gothic universities and aristocratic academies' [28]; it would not be long before a priestly caste would be formed that would be 'more powerful perhaps than that which the people's reason has just overthrown' [29]. Instead of this corporate body, individual initiative would carry truth wherever it would be truly free: 'Render to genius all the latitude of power and liberty that it demands; proclaim its inalienable rights; shower public honours and rewards on all useful interpreters of nature wherever they may be found; do not confine in a narrow circle those intellects (*lumières*) that seek only to cast their light afar' [30]. No organization, just an accorded liberty: 'Those citizens skilled (*éclairés*) in letters and in the arts are invited to take up teaching throughout the French Republic'. No examinations and no qualifications other than age, experience, and the respect of the citizens; whoever wished to teach mathematics, the fine arts, or medicine had only to obtain from his municipality a certificate of integrity and good citizenship: if need be, and if he deserved it, he might also get the local authorities to lend him the materials needed for teaching or experimentation. These lessons, freely given, would be paid for by the pupils by arrangement with the master; but, for deserving cases, the municipality might also give grants. In this regime of economic liberalism and competition, education returned, in a sense, to the freedom of the ancient Greeks: knowledge is spontaneously transmitted by the Word, and the Word that contains most truth prevails. And as if to give a note of nostalgia and inaccessibility to his dream, to lend it a still more Greek stamp that would place his intentions above reproach, the better to conceal his real aims, Fourcroy proposed that after twenty-five years of teaching, the masters should, like so many Socrates recognized at last by a better Athens, be housed and fed throughout their long old age.

Paradoxically, it was the Mountain, and those closest to Robespierre, who defended ideas similar to Condorcet's project. Le Pelletier, whose plan, after its author's assassination, was taken over first by Robespierre, then by Romme (once the Girondists had fallen), who proposed a centralized system of education that would be controlled at every level by the state; even within the Mountain there was concern about these '40,000 bastilles in which it was proposed to imprison the next generation' [31]. Bouquier, a member of the Comité d'Instruction Publique, supported by the Jacobins, proposed a compromise plan that was less archaic than that of the Girondists and less rigid than that of Le Pelletier and Romme. He made an important distinction between 'knowledge that was indispensable to the citizen', without which he could not become a free man—the state owes him this instruction, as it owes him liberty itself—and 'knowledge necessary to society', which the state 'is under an obligation to encourage, but which it can neither organize nor control as it can the former; such knowledge serves the collectivity, it does not form the individual'. Medicine belongs with the arts and sciences. In nine of the country's cities, schools of health would be set up, each with seven teachers (*Instituteurs*); Paris would have fourteen such teachers. Furthermore, 'an officer of health will give lessons in the hospitals reserved for women, children, the insane, and those suffering from venereal diseases'. These teachers would be paid by the state (3,500 francs per annum), and selected by juries drawn from 'the administrators of the district, together with the citizens'[32]. Thus, the public consciousness would find in this system of teaching both its free expression and the utility that it seeks.

With Thermidor, the hospital funds were nationalized, the corporations proscribed, societies and academies abolished, and the University, together with its Faculties and Schools of Medicine, ceased to exist; but the Convention did not have time to implement the policy of assistance that they had accepted in principle, or to lay down limits for the free practice of medicine, or to define what qualifications were necessary to it, or to decide on the form that its teaching should take.

Such difficulties are surprising since for decades each of these questions had been thoroughly discussed, and the wide range of solutions offered certainly revealed a conceptual mastery of the problems; and, above all, since the Legislative Assembly had laid down in principle what, from Thermidor to the Consulate, was to be rediscovered as the solution.

Throughout this whole period, an indispensable structure was lacking: a structure that might have given unity to a form of experience already defined by individual observation, the examination of cases, the everyday practice of diseases, and a form of teaching that everyone knew ought really to be given in the hospital rather than in the Faculty, and in the whole course of the concrete world of disease. What one did not know was how to express in words what one knew to be given only to the gaze. The *Visible* was neither *Dicible* nor *Discible*.

This was because, despite the great changes that had come about in the theories of medicine in the last fifty years and despite the large number of new observations, the subject of medicine remained the same, the position of knowing and perceiving the subject remained the same, and concepts were formed according to the same rules. Or, rather, medical knowledge as a whole obeyed two types of regularity: the first was that of individual, concrete perceptions, mapped out in accordance with the nosological table of morbid species; the second, that of the continuous, over-all, quantitative registration of a medicine of climates and places.

The entire pedagogical and technical reorganization of medicine faltered on account of a central lacuna: the absence of a new, coherent, unitary model for the formation of medical objects, perceptions, and concepts. The political and scientific unity of the medical institution implied, for its realization, this mutation in depth. But, for the reformers of the French Revolution, this unity was effectuated only in the form of theoretical themes that reorganized, after the event, already constituted elements of knowledge.

These fluctuating themes certainly demanded a unity of knowledge and of medical practice; they indicated an ideal place for it; but they were also the principle obstacle to its realization. The idea of a transparent, undivided domain, exposed from top to bottom to a gaze armed nonetheless with its privileges and qualifications, dissipated its own difficulties in the powers accorded to liberty: in liberty, disease was to formulate of itself an unchanging truth, offered, undisturbed, to the doctor's gaze; and society, medically invested, instructed, and supervised, would, by that very fact, free itself from disease. The great myth of the *free gaze*, which, in its fidelity to *discovery* receives the virtue to *destroy*; a purified purifying gaze; which freed from darkness, dissipates darkness. The cosmological values implicit in the *Aufklärung* are still at work here. The medical gaze, whose powers were beginning to be recognized, had not yet been given its technological structure in the clinical organization; it was only one segment of the dialectic of the *Lumières* transported into the doctor's eye.

For reasons that are bound up with the history of modern man, the clinic was to remain, in the opinion of most thinkers, more closely related to the themes of light and liberty—which, in fact, had evaded it—than to the discursive structure in which, in fact, it originated. It is often thought that the clinic originated in that free garden where, by common consent, doctor and patient met, where observation took place, innocent of theories, by the unaided brightness of the gaze, where, from master to disciple, experience was transmitted beneath the level of words. And to the advantage of a historical view that relates the fecundity of the clinic to a scientific, political, and economic *liberalism*, one forgets that for years it was the ideological theme that prevented the organization of clinical medicine.

NOTES

[1] Boissy d'Anglas, *Adresse à la Convention 25 pluviôse an II*. Quoted in Guillaume, *Procés-verbaux du comité d'instruction publique de la Convention* (Vol. II, pp. 640–2).

[2] Bloch and Tutey, *Procés-verbaux et rapports du Comité de Mendicité* (Paris, 1911, p. 395).

[3] *Ibid.*, p. 396.

[4] Asylums in which beggars, the poor, the workless, prostitutes, political agitators, and all those regarded as actual or potential agents of disorder were interned.

[5] Quoted in Imbert, *Le droit hospitalier sous la Révolution et l'Empire* (Paris, 1954, p. 52).

[6] Cabanis, *Du degré de certitude de la médecine* (3rd edn., Paris, 1819, p. 135 and p. 154).

[7] *Ibid.*, p. 146, n. 1.

[8] *Ibid.*, p. 135.

[9] Tenon, *Mémoires sur les hôspitaux* (Paris, 1788, p. 359).

[10] *Ibid.*, p. 354.

[11] J.-B. Duvergier, *Collection complète des lois . . .* , vol. IV, p. 325.

[12] *Archives Parlementaires*, Vol. LVI, p. 646; quoted in Imbert, *Le droit hospitalier sous la Révolution et l'Empire*, p. 76, n. 29.

[13] *Ibid.*, p. 78.
[14] Law of 19 March 1793.
[15] Saint-Just in Buchez and Roux, *Histoire parlementaire*, Vol. XXXV, p. 296.
[16] Articles 1, 6, 9, 10, 14, and 22.
[17] Articles 26 and 27. The complete text of the Marly decrees is quoted by Gilibert, *L'anarchie médicinale* (Neuchâtel, 1772, Vol. II, pp. 58–118).
[18] On this subject, see Gilibert, *op. cit.*; Thiery, *Voeux d'un patriote sur la médecine en France*, 1789; this text would appear to have been written in 1750 and not published until the Estates General.
[19] Cantin, *Projet de réforme adressé à l'Assemblée Nationale* (Paris, 1790, p. 14).
[20] *Ibid.*
[21] Cabanis, *op. cit.*
[22] Jadelot, *Adresse à Nos Seigneurs de l'Assemblée Nationale* (Nancy, 1790, p. 7).
[23] Cf. above, p. 29.
[24] Thiery, *op. cit.*; J.-P. Gallot, *Vues générales sur la restauration de l'art de guérir* (Paris, 1790).
[25] Thiery, *op. cit.*, pp. 89–98.
[26] Cabanis, *Observations sur les hôpitaux* (Paris, 1790, pp. 32–3).
[27] Durand Maillane, J. Guillaume, *Procès-verbaux du Comité d'Instruction publique de la Convention*, Vol. I, p. 124.
[28] Fourcroy, *Rapport sur l'enseignement libre des sciences et des arts* (Paris, Year II, p. 2).
[29] *Ibid.*, p. 2.
[30] *Ibid.*, p. 8.
[31] Sainte-Foy, *Journal de la Montagne*, no. 29, 12 December 1793.
[32] Fourcroy, *op. cit.*

4 · The Old Age of the Clinic

The principle that medical knowledge formed for itself at the very bedside of the patient does not date from the end of the eighteenth century. Many, if not all, the revolutions in medicine have been carried through in the name of this experience, presented as primary source and constant norm. But what was constantly changing was the very grid according to which this experience was given, was articulated into analysable elements, and found a discursive formulation. Not only the names of diseases, not only the grouping of systems were not the same; but the fundamental perceptual codes that were applied to patients' bodies, the field of objects to which observation addressed itself, the surfaces and depths traversed by the doctor's gaze, the whole system of orientation of this gaze also varied.

Medicine had tended, since the eighteenth century, to recount its own history as if the patient's bedside had always been a place of constant, stable experience, in contrast to theories and systems, which had been in perpetual change and masked beneath their speculation the purity of clinical evidence. The theoretical, it was thought, was the element of perpetual change, the starting point of all the historical variations in medical knowledge, the locus of conflicts and disappearances; it was in this theoretical element that medical knowledge marked its fragile relativity. The clinic, on the other hand, was thought to be the element of its positive accumulation: it was this constant gaze upon the patient, this age-old, yet ever renewed attention that enabled medicine not to disappear entirely with each new speculation, but to preserve itself, to assume little by little the figure of a truth that is definitive, if not completed, in short, to develop, below the level of the noisy episodes of its history, in a continuous historicity. In the non-variable of the clinic, medicine, it was thought, had bound truth and time together.

Hence all those somewhat mythical accounts by which, at the end of the eighteenth and the beginning of the nineteenth centuries, the history of medicine was put together. It is in the clinic, it was said, that medicine found its possibility of origin. At the dawn of mankind, prior to every vain belief, every system, medicine in its entirety consisted of an immediate relationship between sickness and that which alleviated it. This relationship was one of instinct and sensibility, rather than of experience; it was established by the individual from himself to himself, before it was caught up in a social network: 'The patient's sensibility tells him whether this or that position makes him more comfortable or torments him' [1]. It is this relationship, established without the mediation of knowledge, that is observed by the healthy man; and this observation itself is not an option for future knowledge; it is not even an act of awareness (*prise de conscience*); it is performed immediately and blindly: 'A secret voice tells us here: contemplate nature' [2]; multiplied by itself, transmitted from one to another, it becomes a general form of consciousness of which each individual is both subject and object: 'Everyone, without distinction, practised this medicine . . . each person's experiences were communicated to others . . . and this knowledge passed from father to children' [3]. Before it became a corpus of knowledge (*un savoir*), the clinic was a universal relationship of mankind with itself: the age of absolute happiness for medicine. And the decline began when writing and secrecy were introduced, that is, the concentration of this knowledge in a privileged group, and the dissociation of the immediate relationship, which had neither obstacle nor limits between Gaze and Speech (*Parole*): what was known was no longer communicated to others and put to practical use once it had passed through the esotericism of knowledge [4].

No doubt medical experience remained open for a long time, and succeeded in striking a balance between *seeing* and *knowing* (*le voir et le savoir*) that protected it from error: 'In far-off times, the art of medicine was taught in the presence of its object and young men learnt medical science at the patient's bedside'; the patients were often accommodated in the doctor's own house, and the pupils accompanied their masters at all hours on the rounds of their patients [5]. Hippocrates seems to be both the last witness and the most ambiguous representative of this balance: fifth-century Greek medicine would seem to be no more than the codification of this universal, yet immediate, clinical medicine; it formed the first total consciousness of this clinical medicine, and in this sense, it seems to be as 'pure and simple' [6] as that first experience; but insofar as it organizes it into a systematic corpus in order to facilitate and shorten the study of it a new dimension is introduced into medical experience: that of a corpus of knowledge that can be said to be, quite literally, blind, since it has no gaze. This unseeing knowledge is at the source of illusion; a medicine haunted by metaphysics becomes possible: 'When Hippocrates had reduced medicine to a system, observation was abandoned and philosophy was introduced into medicine' [7].

Such is the occultation that has made possible the long history of systems, with 'the multiplicity of different sects opposing and contradicting one another' [8]. A history, therefore, that negates itself, preserving from time only its destructive mark. But beneath that destructive history lies another history, one more faithful to time because closer to its original truth. Into this history is imperceptibly gathered the silent life of the clinic. It remains beneath all 'speculative theories' [9], keeping medical practice in contact with the world of perception, and opening it up to the immediate landscape of truth: 'There have always been doctors who, with the help of that analysis that comes so naturally to the human mind, having deduced from the patient's appearance all the data needed concerning his idiosyncrasy, have been content simply to study the symptoms . . .' [10]. Immobile, but always close to things, the clinic gives medicine its true historical movement, it effaces systems, while the experience that contradicts them accumulates its truth. Thus a fruitful continuity is found that guarantees to pathology 'the uninterrupted uniformity of that science throughout the centuries' [11]. Over and against systems, which belong to negating time, the clinic is the positive time of knowledge. It is not to be invented, therefore, but to be rediscovered: it was already there with the first forms of medicine; it has constituted its full plenitude; it is enough therefore to deny that which denies it, to destroy that which in relation to it is nothingness—that is, 'the prestige' of systems—and to leave it at last 'to enjoy its full rights' [12]. Medicine would then be on a level with its truth.

This ideal account, which is to be found so frequently at the end of the eighteenth century, must be understood in relation to the recent establishment of clinical institutions and methods. It presented them as the restitution of an eternal truth in a continuous historical development in which events alone have been of a negative order: oblivion, illusion, concealment. In fact, this way of rewriting history itself evaded a much truer but much more complex history. It masked that other history by assimilating to clinical method all study of cases, in the old sense of the word; and, therefore, it authorized all subsequent simplifications whereby clinical medicine became simply the examination of an individual.

In order to understand the meaning and structure of clinical experience, we must first rewrite the history of the institutions in which its organizational effort has been manifested. Up to the last years of the eighteenth century, this history, as a chronological succession, is extremely thin.

In 1658, François de la Boe opened a clinical school in the hospital at Leyden; he published the resulting observations under the title of *Collegium Nosocomium* [13]. The most illustrious of his successors was Boerhaave. It is also possible that there was a chair of clinical medicine at Padua from the end of the sixteenth century [14]. In any case, it was at Leyden that the practice began, with Boerhaave and his pupils, in the eighteenth century, of setting up chairs or institutes of clinical medicine. In 1720, some of Boerhaave's pupils reformed the University of Edinburgh and set up a teaching hospital on the Leyden model; their example was followed in London, Oxford, Cambridge, and Dublin [15]. In 1733, Van Swieten was asked to submit plans for the establishment of a clinic at the hospital of Vienna: the first holder of the chair was de Haen, a pupil of Boerhaave's, and he was succeeded first by Stoll, then by Hildenbrand [16]; this example was followed at Göttingen, where Brendel, Vogel, Baldinger, and J.-P. Franck taught in turn [17]. At Padua, a number of hospital beds were devoted to clinical medicine, with Knips as professor; Tissot, who was appointed to set up a clinic at Pavia, explained the broad outlines of his plans in his inaugural lecture on 26 November 1781 [18]. About 1770, Lacassaigne, Bourru, Guilbert, and Colombier had wanted to organize privately and at their own expense a small, twelve-bed hospital for acute cases, in which the doctors would combine the teaching of practical medicine with the treatment of the patients [19]; but the project

failed. The Faculty, and the medical profession in general, were too
concerned to maintain the old state of affairs whereby practical
teaching was given individually, privately, and with great expense
of time and money by the more celebrated consultants. It was in the
military hospitals that clinical teaching was first organized; the
Règlement pour les hôpitaux, drawn up in 1775, states in its article
XIII that each year of study must include 'a course of practical and
clinical medicine of the principal diseases to be found among the
troops in the armies and garrisons' [20]. And Cabanis quotes as an
example the clinic attached to the naval hospital at Brest founded
by Dubreil under the auspices of the Maréchal de Castries [21]. To
conclude, one might mention the setting up of a maternity clinic in
Copenhagen in 1787 [22].

Such, it seems, are the facts. In order to understand their meaning and disentangle the problems that they pose, one must first re-
examine a number of observations that should diminish their
importance. The examination of cases, the writing up of detailed
accounts of them, and their relationship with a possible explanation
belong to an essential tradition that has never been in question in
medical experience; the organization of the clinic is not correlative
with the discovery of the individual fact in medicine; the innumerable collections of cases published since the Renaissance is proof
enough of this. Furthermore, there was also a very wide recognition
of the need for teaching through practice itself: hospital visits by
apprentice doctors was now widespread; and some of these apprentice doctors would complete their training in a hospital in
which they lived and practised under the supervision of a doctor
[23]. What, therefore, was so new and so important about those
clinical establishments that the eighteenth century, especially
towards its close, valued so highly? In what respect could this
proto-clinic be distinguished from the spontaneous practice that had
once been synonymous with medicine, on the one hand, and the
clinic as it was later to become organized into a complex, coherent
corpus combining a form of experience, a method of analysis, and
a type of teaching, on the other? Can it be attributed to a specific
structure that might be regarded as peculiar to the eighteenth-century medical experience with which it is contemporary?

1. This proto-clinic is more than a successive, collective study
of cases: it must gather together and make perceptible the organized
corpus of nosology. The clinic, therefore, could be neither *open to
all*, as a doctor's daily practice can be, nor *specialized*, as it was to

become in the nineteenth century: it was neither the enclosed domain of what one has chosen to study nor the open statistical field of what one cannot but receive; it is enclosed upon the didactic totality of an ideal experience. Its task is not to indicate individual cases, with their dramatic points and their particular characteristics, but to manifest the complete circle of diseases. The Edinburgh clinic was for long a model of its kind; it was organized in such a way that 'those cases that seem most instructive' could be brought together [24]. Before being a meeting of patient and doctor, a truth to be deciphered and an ignorance, and in order to be such a meeting, *the clinic must form, constitutionally, a structured nosological field.*

2. Its contact with the hospital was of a special kind. It was not the direct expression of the hospital, since a principle of choice serves as a selective limit between them. This selection is not simply quantitative, though, according to Tissot, the number of beds should not exceed thirty [25]; it is not only qualitative, though it tends to prefer those cases that have a high instructive value. By operating a process of selection, it alters in its very nature the way in which the disease is manifested, and the relationship between the disease and the patient; in the hospital one is dealing with individuals who happen to be suffering from one disease or another; the role of the hospital doctor is to discover the disease in the patient; and this interiority of the disease means that it is often buried in the patient, concealed within him like a cryptogram. In the clinic, on the other hand, one is dealing with diseases that happen to be afflicting this or that patient: what is present is the disease itself, in the body that is appropriate to it, which is not that of the patient, but that of its truth. It is 'the different diseases that serve as the text' [26]: the patient is only that through which the text can be read, in what is sometimes a complicated and confusing state. In the hospital, the patient is the *subject* of his disease, that is, he is a *case*; in the clinic, where one is dealing only with *examples*, the patient is the accident of his disease, the transitory object that it happens to have seized upon.

3. The clinic knows its truth, therefore, only in its synthetic form. It is already completely given in that form, and its manifestations are no more than its consequences. In this form of teaching, the pupil may well not possess the key from the outset. Tissot is in favor of making him look for it for a long time. He suggests that each patient in the clinic should be entrusted to two students;

they and they alone would examine him, 'with decency, with gentleness, and with that kindness that is so consoling for those poor unfortunates' [27]. They would begin by questioning him as to his country of origin, the constitutions that are common there, his profession, his previous illnesses, the way in which his present illness began, the remedies already taken; they would investigate his vital functions (breathing, pulse, temperature), his natural functions (senses, faculties, sleep, pain); they would also have to 'palpate the abdomen in order to ascertain the state of his viscera' [28]. But what are they looking for, and what hermeneutic principle should guide them in their examination? What are the relations set up between the phenomena observed, the antecedences ascertained, the disorders and deficiencies noted? Nothing more than will enable one to name the disease. Once the designation has been carried out, it will be an easy matter to deduce the causes, the prognosis, and the indications, by 'asking oneself: What is wrong with this patient? What is to be put right?' [29]. Compared with later methods of examination, that recommended by Tissot is hardly less meticulous, apart from a few details. The difference between this investigation and the 'clinical examination' lies in the fact that in the former no inventory of a sick organism is made; one retains those elements that enable one to put one's hand on an ideal key—a key that has four functions, since it is a mode of designation, a principle of coherence, a law of evolution, and a body of precepts. In other words, the gaze that traverses a sick body attains the truth that it seeks only by passing through the dogmatic stage of the *name*, in which a double truth is contained: the hidden, but already present truth of the disease and the enclosed truth that is clearly deducible from the outcome and from the means. So it is not the gaze itself that has the power of analysis and synthesis, but the synthetic truth of language, which is added from the outside, as a reward for the vigilant gaze of the student. In this clinical method, in which the density (*épaisseur*) of the perceived hides only the imperious and laconic truth that names, it is a question not of an *examination*, but of a *deciphering*.

4. So it is understandable that the clinic should have had only one direction—from top to bottom, from constituted knowledge to ignorance. In the eighteenth century, there were only teaching clinics, though only in a limited form, since it was not conceded that the doctor should be able by this method at any moment to read the truth that nature had deposited in the illness. The clinic was

concerned only with the instruction, in the narrow sense of the word, that is given by a master to his pupils. It was not in itself an experience, but a condensed version, for the use of others, of previous experience. 'The professor indicates to his pupils the order in which objects must be observed in order to be seen and remembered more easily' [30]. In no sense was the clinic to *discover* by means of the gaze; it merely duplicated the art of demonstrating (*démontrer*) by showing (*montrer*). This was how Desault understood the lessons of clinical surgery that he gave at the Hôtel-Dieu from 1781 onwards:

> under the eyes of his listeners, he brought in the most seriously sick patients, classified their disease, analysed its features, outlined the action that was to be taken, carried out the necessary operations, gave an account of his methods and the reasons for them, explained each day the changes that had occurred, and then presented the state of the cured patients . . . or demonstrated on the lifeless body the alterations that had rendered further exercise of his art useless [31].

5. The example of Desault shows, however, that this speech (*parole*), didactic in essence as it may be, accepted in spite of everything the judgement and risk of the future. In the eighteenth century, the clinic was not a structure of medical experience, but it was experience at least in the sense that it was a test—a test of knowledge that time must confirm, a test of prescriptions that will be proved right or wrong by the outcome, before the spontaneous jury of students: there is a sort of contest, before witnesses, with the disease, which has its own word to say, and which, despite the dogmatic speech used to designate it, possesses its own language. Thus the lesson given by the master may turn against him, and provide, despite his vain language, a lesson that belongs to nature itself. Cabanis explains the lesson to be drawn from a bad lesson in this way: if the professor makes a mistake, 'his failures are soon unmasked by nature . . . whose language can be neither stifled nor altered. They may even prove to be more useful than his successes, and render more ineffective images which might otherwise have made only a slight impression on them' [32]. It is when the master's designation fails, therefore, and when time has proved its worthlessness, that the movement of nature is recognized for itself: the language of knowledge remains silent, and one observes. This test showed great honesty, for it was linked to its proper stake according to a sort of contract renewed daily. At the Edinburgh clinic the

students kept a record of the diagnosis made, of the state of the patient at every visit, and of the medicines taken during the day [33]. Tissot, who also recommended the keeping of a diary, adds in his report to Count Firmian, in which he describes the ideal clinic, that these diaries should be published each year [34]. Finally, in fatal cases, dissection must provide a last confirmation [35]. Thus the synthetic, designating speech of knowledge is confronted by the audible language of nature in a chronicle of observations that form a mixed syntax—a sort of neutral, arbitrary language. But, in fact, the eighteenth century failed to give this language a status, a coherent grammar. It was not yet a scientific language, but only a 'gaming' language (*un langage de jeu*); truth did not find its original formulation in that language; it ran the risk, according to the play of chance or skill, of winning or losing.

In the eighteenth century, then, the clinic was already a much more complex form than a mere knowledge of cases. And yet, it did not prove to be of great value in the actual movement of scientific knowledge; it formed a marginal structure that was articulated upon the hospital field without having the same configuration; it was intended as a means of teaching medical practice, which it symbolized rather than analysed; it grouped all experience around the play of a verbal unmasking that was not simply its form of transmission, theatrically retarded.

But in a few years, the last years of the century, the clinic was to undergo a sudden, radical restructuring: detached from the theoretical context in which it was born, it was to be given a field of application that was no longer confined to that in which knowledge was *said*, but which was co-extensive with that in which it was born, put to the test, and fulfilled itself: it was to be identified with the *whole* of medical experience. For this, it had to be armed with new powers, detached from the language on the basis of which it had been offered as a lesson, and freed for the movement of discovery.

NOTES

[1] Cantin, *Projet de réforme adressé à l'Assemblée Nationale* (Paris, 1790, p. 8).
[2] *Ibid.*

[3] Coakley Lettson, *Histoire de l'origine de la médecine* (Fr. trans., Paris, 1778, p. 7).
[4] *Ibid.*, pp. 9–10.
[5] P. Moscati, *De l'emploi des systèmes dans la médecine pratique* (Fr. trans., Strasbourg, Year VII, p. 13).
[6] P.-A.-O. Mahon, *Histoire de la médecine clinique* (Paris, Year XII, p. 323).
[7] Moscati, *op cit.*, pp. 4–5.
[8] *Ibid.*, p. 26.
[9] Dezeimeris, *Dictionnaire historique de la médecine* (Paris, 1828, vol. I), article 'Clinique', pp. 830–7.
[10] J.-B. Regnault, *Considérations sur l'État de la médecine* (Paris, 1819, p. 10).
[11] P.-A.-O. Mahon, *op. cit.*, p. 324.
[12] *Ibid.*, p. 323.
[13] Leyden, 1667.
[14] Cf. Comparetti, *Saggio delle scuola clinica nolle spidle di Padova.*
[15] J. Aikin, *Observations sur les hôpitaux* (Fr. trans., Paris, 1777, pp. 94–5).
[16] A. Störck, *Instituta Facultatis mediae Vivobonensis* (Vienna, 1775).
[17] Dezeimeris, *op. cit.*
[18] Tissot, *Essai sur les études de médecine* (Lausanne, 1785, p. 118).
[19] Colombier, *Code de Justice militaire*, II, pp. 146–7.
[20] 'Règlement pour les hôpitaux militaires de Strasbourg, Metz et Lille, fait sur l'ordre du roi par P. Haudesierck, 1775,' quoted by Boulin, *Mémoires pour servir à l'histoire de la médecine* (Paris, 1776, vol. II, pp. 73–80).
[21] Cabanis, *Observations sur les hôpitaux* (Paris, 1790, p. 31).
[22] J.-B. Demangeon, *Tableau historique d'un triple établissement réuni en un seul hospice à Copenhague* (Paris, Year VII).
[23] This was the case in France, at the Hôpital Général, for example; throughout the eighteenth century an apprentice surgeon lived at La Salpêtrière, accompanied the surgeon on his visits, and administered a little simple treatment himself.
[24] Aikin, *op. cit.*
[25] Tissot, 'Mémoire pour la construction d'un hôpital clinique', *op. cit.*
[26] Cabanis, *op. cit.*, p. 30.
[27] Tissot, *op. cit.*, p. 120.
[28] *Ibid.*, pp. 121–3.
[29] *Ibid.*, p. 124.
[30] Cabanis, *op. cit.*, p. 30.
[31] M.-A. Petit, 'Éloge de Desault', *Médecine du Coeur*, p. 108.
[32] Cabanis, *op. cit.*, p. 30.
[33] J. Aikin, *op. cit.*, p. 95.
[34] Tissot, *op. cit.*
[35] Cf. *ibid.*, and M.-A. Petit, *op. cit.*

5 · The Lesson of the Hospitals

In the article entitled 'Abus' in the *Dictionnaire de Médecine*, Vicq d'Azyr sees the organization of a system of teaching within the hospital as the universal solution for the problems of medical training; that, for him, is the major reform to be carried out: 'Diseases and death offer great lessons in hospitals. Are we benefiting from them? Are we writing the history of the illnesses that strike so many victims in our hospitals? Do we teach in our hospitals the art of observing and treating diseases? Have we set up any chairs of clinical medicine in our hospitals?' [1] Yet, in a very short time, this reform of the teaching system was to assume a much wider signficance; it was recognized that it could reorganize the whole of medical knowledge and establish, in the knowledge of disease itself, unknown or forgotten, but more fundamental, more decisive forms of experience: the clinic and the clinic alone was capable of 'reviving among the moderns the temples of Apollo and Aesculapius' [2]. A way of teaching and *saying* became a way of learning and *seeing*.

At the end of the eighteenth century, as at the beginning of the Renaissance, education was given a positive value as enlightenment: to train was a way of bringing to light, and therefore of discovering. The childhood and youth of things and men were endowed with an ambiguous power: to tell of the birth of truth; but also to put to the test the tardy truth of men, to rectify it, to bring it closer to its nudity. The child became the immediate master of the adult insofar as true education was identified with the very genesis of truth. In every child things tirelessly repeat their youth, the world resumes contact with its native form: he who looks for the first time is never an adult. When it has untied its old kinships, the eye is able to open at the unchanging, ever-present level of things; and of all the senses and all sources of knowledge (*tous les savoirs*), it is intelligent enough to be the most unintelligent by repeating so skilfully its distant ignorance. The ear has its preferences, the hand its lines and its folds; the eye, which is akin to light, supports only the present. What allows man to resume contact with childhood and to rediscover the permanent birth of truth is this bright, distant, open naïvety of the gaze. Hence the two great mythical experiences on which the philosophy of the eighteenth century had wished to base its beginning: the foreign spectator in an unknown country, and the man born blind restored to light. But Pestalozzi and the *Bildungsromane* also belong to the great theme of Childhood-Gaze. The discourse of the world passes through open eyes, eyes open at every instant as for the first time.

With the reaction that set in after 9 Thermidor, the pessimism of Cabanis and Cantin seemed to be confirmed: the expected 'brigandage' became widespread [3]. From the beginning of the war, but especially after the mass rising of autumn 1793, many doctors joined the army, either as volunteers or conscripts; the quacks had a free field [4]. A petition addressed on 26 Brumaire Year II to the Convention and drawn up by a certain Caron, of the Poissonnière section, was still denouncing doctors trained by the Faculty as vulgar 'charlatans' against whom the people wished to be defended [5]. But this fear soon took on a different shape, and the danger was seen to come from the real charlatans who were not doctors: 'The public has been subjected to a host of ill-taught individuals who, on no other authority but their own, have set themselves up as masters of the art, who hand out remedies quite indiscriminately and threaten the lives of several thousand citizens' [6]. The disasters caused by this 'savage' medicine were so great in one department (Eure) that the Directoire, alerted to the danger, recalled the Assemblée des Cinq-Cents [7] and, on two occasions, the 13 Messidor Year IV and the 24 Nivôse Year VI, the government requested the legislature to limit this dangerous liberty: 'O representative citizens, the nation is making its maternal cries heard

and the executive Directory is their organ! This is certainly a matter of the utmost urgency: the delay of a single day may mean the
death of several citizens' [8]. Inadequately trained doctors and
experienced quacks were equally dangerous, especially when the
hospitalization of the poor and sick became increasingly difficult.
The nationalization of hospital funds sometimes went so far as the
confiscation of liquid capital, and many bursars had no other course.
but to turn out boarders whom they could no longer keep. Sick or
wounded soldiers occupied many of the establishments, and the
municipalities were delighted that they no longer had to find the
resources for their hospitals: at Poitiers, on 15 July 1793, 200 patients were turned out of the Hôtel-Dieu to make room for
wounded soldiers whose board was paid for by the army [9]. This
dehospitalization of illness, brought about by a spontaneous convergence of hard facts and revolutionary dreams, far from restoring
pathological essences to a truth of nature, and reducing them by
that very fact, merely added to the ravages that they were already
causing and left the population without either protection or help.

No doubt many medical officers came and settled down as
civilian practitioners in town and country on leaving the army at
the end of the Thermidorian period and the beginning of the
Directoire. But the quality of these doctors was not uniform.

Many medical officers were very lacking in training and experience. In Year II, the Comité de Salut Public had asked the
Comité d'Instruction Publique to draft a bill whereby 'officers of
health can be trained without delay for the needs of the armies of
the Republic' [10]; but the situation had been too urgent, all volunteers had been accepted and given a rapid training, and apart from
the first-grade officers of health, who had to show proof of previous
training, they had no further knowledge of medicine than what they
had just been taught. Even in the army, these ill-trained practitioners
had been criticized for their numerous mistakes [11]. But when
they practised among the civilian population, without the supervision of their seniors, such doctors caused far worse damage; there
was the case of an officer of health in the Creuse who killed his
patients by administering purges of arsenic [12]. From all sides demands flowed in for proper control and supervision and for new
legislation: 'With how many ignorant murderers will you inundate
France if you authorize second- and third-class physicians, surgeons, and chemists ... to practise their respective professions without a new examination; ... it is, above all, in that homicidal Society

that one still finds the most respected, most dangerous charlatans, those whom the law must make it its task to supervise' [13].

Protective bodies sprang up spontaneously against this state of affairs. Some of the more precariously based were popular in origin. If certain of the more moderate Parisian sections remained faithful to the axiom of the Mountain—'No more indigents, no more hospitals'—and continued to demand the distribution of individual aid, to benefit the sick who were cared for at home [14], others, including the poorest, were forced, by penury and the difficulty of obtaining treatment, to demand the setting up of hospitals in which the poor and sick would be lodged, fed, and treated; they hoped for a return to the principle of the poorhouse [15]; and houses were opened, clearly without governmental initiative, with funds raised by popular societies and assemblies [16]. After Thermidor, on the other hand, the movement came from above. The enlightened classes, the intellectual circles, who had returned to power or obtained it at last, wished to restore to knowledge the privileges that would be able to protect both the social order and individual lives. In several cities, the administrations, 'affrighted by the ills that they had witnessed' and 'afflicted by the silence of the law', did not wait until the legislature had made its decisions: they decided to establish their own control over those who claimed to practise medicine; they set up commissions composed of doctors of the Ancien Régime, who would pass judgement on the qualifications, knowledge, and experience of all newcomers [17]. Furthermore, certain Faculties that had been closed down continued to function in semi-secrecy: the former professors gathered around themselves those who wished to learn, and were accompanied by these students on their visits; if they were placed in charge of a hospital department, it was there, at their patients' bedside, that they gave their teaching and were able to judge the aptitude of their pupils. Sometimes, when these purely private studies were completed, the professors even issued a sort of unofficial diploma, certifying that the holder had become a true doctor.

Montpellier provided what was no doubt a fairly rare example of a meeting place for these various forms of reaction: one can see the appearance there of the need to train doctors for the army, the use of the old medical qualifications sanctified by the Ancien Régime, the intervention of popular assemblies and of the local administration, and the spontaneous beginnings of clinical experience. Baumes, a former university professor, had been appointed, both for his experience and his republican opinions, to the military hospital at Saint-Éloi. There he was to make a selection from among the candidates for the posts of officers of health; but since no teaching had been organized, the medical students appealed to the 'society of the people' (*société populaire*), which, by means of a petition, persuaded the district administration to establish clinical teaching at the hospital of Saint-Éloi, under Baumes's supervision. In the following year, 1794, Baumes published the results of his observations and teaching: 'Method for curing diseases as they appear in the course of the medical year' [18].

This may be a privileged example, but it is no less significant for that. By a spontaneous convergence of pressures and demands proceeding from social classes, institutional structures, technological or scientific problems of very different kinds, an experience was beginning to be formed by a kind of orthogenesis. To all appearances, it was simply reviving, as the only possible way of salvation, the clinical tradition that had been developed in the eighteenth century. In fact, what was involved was something quite different. In that autonomous movement and the quasi-clandestinity that abetted and protected it, this return to the clinic was in fact the first organization of a medical field that was at once composite and fundamental: composite because, in its everyday practice, hospital experience resembles the general form of a pedagogic system; but fundamental, too, because, unlike the eighteenth-century clinic, it is not a question of an encounter, after the event, of a previously formed experience and an ignorance to be dissipated. It is a question, in the absence of any previous structure, of a domain in which truth teaches itself, and, in exactly the same way, offers itself to the gaze of both the experienced observer and the naïve apprentice; for both, there is only one language: the hospital, in which the series of patients examined is itself a school. The abolition of both the old hospital structures and the university made possible, then, the immediate communication of teaching within the concrete field of experience; furthermore, it effaced dogmatic language as an essential stage in the transmission of truth. The silencing of university speech (*la parole universitaire*) and the abolition of the professorial chair made it possible, beneath the old language, in the obscurity of a partly blind practice, driven this way and that by circumstances, for a language without words, possessing an entirely new syntax, to be formed: a language that did not owe its truth to speech but to the gaze alone. In this hasty recourse to the clinic, another clinic, with an entirely new configuration, was born.

It is hardly surprising if suddenly, at the end of the Convention, the theme of an entirely new medicine, based upon the clinic, swept away the theme of a medicine restored to liberty that had been dominant right up to 1793. What occurred was neither reaction (although the social consequences were, in general, 'reactionary'), nor progress (although medicine, as a practice and as a science, benefitted in several ways); what occurred was the restructuring, in a precise historical context, of the theme of 'medicine in liberty': in a liberated domain, the necessity of the truth that communicated itself to the gaze was to define its own institutional and scientific structures. It was not only out of political opportunism, but no doubt also out of an obscure fidelity to coherences that no twisting in events could deflect, that in Year II the same Fourcroy opposed any project aimed at restoring 'the Gothic universities and aristocratic academies' [19] and in Year III demanded that the temporary closure of the Faculties should be used to bring about their 'reform and improvement' [20]; 'murderous quackery and ambitious ignorance' must not be allowed 'to lay their traps for credulous suffering' [21]. What hitherto had been lacking, 'the very practice of the art, the observation of patients in their beds', was to become the essential part of the new medicine.

Thermidor and the Directoire took the clinic as their major theme in the institutional reorganization of medicine: for them, it was a means of putting an end to the dangerous experiment of total liberty, and yet a way of giving it a positive meaning, a way, too, of restoring, as many wished, some of the structures of the Ancien Régime.

I. THE MEASURES OF 14 FRIMAIRE, YEAR III

Fourcroy had been given the task of presenting a report to the Convention on the establishment of an École de Santé in Paris. The justifications that he offered are worth noting, especially in view of the fact that they were taken up virtually *in toto* in the preamble of the law that was in fact passed, though he departs more than once from the letter and the spirit of the project. What was proposed was the establishment, above all, of a single school for the whole of France, modelled on the École Centrale des Travaux Publics, where officers of health would be trained to staff the hospitals, especially the military hospitals: had not 600 doctors been killed in the army in under eighteen months? Apart from the urgency of the situation, and the need to put an end to the malpractices of charlatans, there was a need to remove a number of important objections that might be raised against a measure that ran the risk of restoring the old corporations and their privileges: medicine is a practical science whose truth and success are of interest to the whole nation; by setting up a school, one is not favouring a small handful of individuals, but, through qualified intermediaries, one is helping the people to feel the benefits of truth. As the writer of the report rather awkwardly puts it, 'It is to give fresh life to the several channels that circulate the industrious activity of the arts and sciences through all the ramifications of the social body' [22]. What makes medicine, thus understood, a corpus of knowledge of use to all citizens is its immediate relationship with nature: instead of being, like the old Faculty, the locus of an esoteric, bookish corpus of knowledge, the new school would be 'the temple of nature'; there one would learn not what the old masters thought they knew, but that form of truth open to all that is manifested in everyday practice: 'Practice will be linked to theoretical precepts. Pupils will be practised in chemical experiments, anatomical dissections, surgical operations, and in the use of machinery. Read little, see much, and do much.' They will learn as they practise, at the patient's bedside: instead of useless physiologies, they will learn the true 'art of curing' [23].

The clinic figures, then, as a structure that is essential to the scientific coherence and also to the social utility and political purity of the new medical organization. It represents the truth of that organization in guaranteed liberty. Fourcroy proposed that in three hospitals (the Hospice de l'Humanité, the Hospice de l'Unité, and the Hôpital de l'École), the clinical teaching should be entrusted to professors who would be sufficiently well paid to be able to devote themselves to the task entirely [24]. The public would be freely admitted to the new school of health: in this way, it was hoped that all those who practised medicine without proper training would come of their own free will to complete their experience. In any case, in each district pupils would be chosen who had shown 'good conduct, pure morals, love of the Republic, and a hatred of tyrants, sufficient education, and, above all, a knowledge of some of the sciences that might serve as a preliminary to the art of curing', and they would be sent to the École Centrale de Médecine, to be trained over a period of three years as officers of health [25].

For the provinces, Fourcroy proposed that there should be only special schools. The deputies from the south of France objected, and insisted that Montpellier should also have its École Centrale. Then Ehrman demanded the same privilege for Strasbourg, with the result that the law of 14 Frimaire Year III provided for the setting up of three schools of medicine, each providing a course of teaching lasting three years. In Paris, the 'beginners' class' would study anatomy, physiology, and medical chemistry during the first semester, and materia medica, botany, and physics during the second; throughout the year the students would visit hospitals 'in order to get used to seeing the sick and how they are treated' [26]. Second-year students would study first anatomy, physiology, chemistry, pharmacy, and surgery, then materia medica, internal and external pathology; during this second year, students might be employed in the hospitals 'in the service of the sick'. In the final year, the students would revise what they had learnt in the first two years and, benefitting from the hospital experience already gained, begin their real clinical training. The students would be distributed among three hospitals, in each of which they would remain four months, then move on. The clinical training consisted of two parts: 'The professor would pause at the bedside of each patient long enough to question him and examine him properly; he would draw the students' attention to the diagnostic signs and the important symptoms of the disease'; then, in the lecture hall, the professor would take up the general history of the illnesses observed in the hospital ward, and he would point out their 'known, probable, and hidden' causes, make a prognosis, and provide 'vital', 'curative', or 'palliative' indications [27].

What characterized this reform was not only that the balance of medicine was shifted further in the direction of the clinic, but that this was also counterbalanced by much broader theoretical teaching. Once one defined a practical experiment carried out on the patient himself, one insisted on the need to relate particular knowledge to an encyclopaedic whole. The first two principles by which the new Paris school commented on the law of 14 Frimaire

required that it 'know animal economy from the elementary structure of the inanimate body to the most composite phenomena of the
organism and life' and that it strive to show the relationships that
exist between living bodies and those in nature [28]. Furthermore,
this broadening of interest would bring medicine into contact with
a whole series of problems and practical requirements: by revealing
the inseparableness of the human being with the material conditions
of existence, it would show how 'one can preserve an individual
life as free from ills as much and for as long as men can reasonably
expect'; and it would represent 'the point of contact between the
art of healing and the civil order' [29]. Clinical medicine is not,
therefore, a medicine concerned only with the first degree of
empiricism, seeking to reduce, by some kind of methodical scepticism, all its knowledge and teaching to observation of the visible
alone. At this first stage, medicine is not defined as clinical unless
it is also defined as encyclopaedic knowledge of nature and knowledge of man in society.

II. REFORMS AND CONTROVERSIES IN YEARS V AND VI

The measures passed on 14 Frimaire fell far short of solving all the
problems that presented themselves. By opening the Écoles de
Santé to the public, it was hoped that inadequately trained officers
of health would be attracted, and that by free competition quacks
and amateurs would disappear. Nothing of the kind occurred: the
inadequate number of schools, and the absence of examinations, except for students with scholarships, prevented the formation of a
body of qualified doctors: on four occasions, 13 Messidor Year IV,
22 Brumaire and 4 Frimaire Year V, and 24 Nivôse Year VI, the
Directoire had to remind the Assemblies of the damage caused by
the free practice of medicine, the bad training of practitioners, and
the lack of effective legislation. What was needed, then, was to
find a system for controlling doctors who had set up in practice
since the Revolution, and to extend the recruitment, the rigour, and
the influence of the new schools.

Moreover, the teaching provided by the schools themselves was
open to criticism. The programme was far too broad and too ambitious for a course that lasted for only three years, as it did under
the Ancien Régime: 'By demanding too much, one achieves
nothing' [30]. There was little unity between the various courses:
at the École de Paris, for example, a clinical medicine of symptoms

was taught, while at the same time, in internal pathology, Doublet was teaching the most traditional kind of medicine of species (first, the most general causes, then 'the general phenomena, the nature and character of each class of diseases and of its principal divisions'; he repeated 'the same examination on the genera and species') [31]. But the clinic itself did not provide the training that had been expected of it: there were too many students and too many patients. 'One moves rapidly round the ward, one says a few words about the outcome of this or that development, and then one hastily withdraws, and that is what passes for teaching in a clinic. In the larger hospitals, one usually sees a great many patients, but very few diseases' [32].

Finally, taking full advantage of all these criticisms, the former members of the medical societies were successful in demanding the restoration of a medical profession defined by qualifications and protected by laws: the medical societies, which had disappeared, together with the University, in August 1792, were reconstituted shortly after the passing of the law of 14 Frimaire. The first of these was the Société de Santé, founded on 2 Germinal Year IV with Desgenettes, Lafisse, Bertrand Pelletier, and Leveillé; in principle, it was intended to serve only as a free, neutral organ of information: rapid communication of observations and experiments, knowledge available to all those concerned with the art of healing, in short, a sort of great clinic on a national scale, which would do no more than observe and practise. The society's first prospectus declared:

Medicine rests on precepts for which experience alone can provide the basis. In order to collect them, we need the co-operation of observers. For this reason, several branches of medicine have declined since the destruction of the learned societies. But from now on they will grow and flourish once again under the auspices of a constituted government that cannot but view with satisfaction the formation of free societies of observer-practitioners [33].

It was in this spirit that the society, convinced 'that the isolation of persons ... is entirely prejudicial to the interests of mankind' [34], published a Recueil périodique, which was soon supplemented by another devoted to foreign medical literature. But before long, this universal concern for information revealed what was no doubt its true preoccupation: to regroup those doctors whose competence had been validated by ordinary studies, and to

militate in favour of a new definition of the limits of the free
practice of medicine: 'Let me not be permitted to conceal from
history the memory of those disastrous times when an impious and
barbarous hand smashed in France the altars devoted to the cult of
medicine! They have disappeared, those bodies whose ancient fame
attested to their long-standing successes' [35]. With this selective
rather than informative character, the movement spread to the
provinces: societies were founded at Lyons, Brussels, Nancy,
Bordeaux, and Grenoble. On 5 Messidor of the same year, another
society held its inaugural meeting in Paris, with Alibert, Bichat,
Bretonneau, Cabanis, Desgenettes, Dupuytren, Fourcroy, Larrey,
and Pinel. To a far greater degree than the Société de Santé, it
represented the opinions of the new medicine: the temple gates
must be shut against those who have entered without deserving to,
taking advantage of the fact that 'at the first signal of the Revolution the sanctuary of medicine, like the temple of Janus, was
flung wide open to admit the onrushing crowd' [36]. But the
method of teaching practised in the schools set up in Year III must
also be reformed: a hasty, composite training that provides the doctor with no reliable method of observation and diagnosis; so 'the
philosophical, reasoned march of method must replace the irregular,
tottering walk of unmethodical activity' [37]. In the eyes of public
opinion, outside the Directoire and the Assemblies but not without
their at least tacit assent, and with the constant support of the re-
presentatives of the enlightened bourgeoisie and the idealogues
close to the government [38], the clinical idea assumed a rather
different meaning from that introduced by the legislators of the
Year III.

Article 356 of the Directoire Constitution declared that 'the law
supervises those professions concerned with the health of citizens';
it was on the strength of this article, which seemed to promise
control, limitations, and guarantees, that all the polemics were
conducted. This is not the place to give a detailed account of these
polemics, but the controversy was centered mainly around the question as to whether one should first reorganize the system of teaching, then draw up the conditions for the practice of medicine, or, on
the contrary, first purge the medical body, define the norms of
practice, and only then decide what form medical studies should
take. Between these two theses, the political division was clear-cut;
those least removed from conventional tradition, such as Daunou or

Prieur de la Côte-d'Or, wanted to reintegrate the officers of health and all the amateur practitioners of medicine by providing a very open system of teaching; the others, around Cabanis and Pastoret, wanted to hasten the reconstitution of an enclosed medical body. At the beginning of the Directoire, it was the first group that had most support.

The first plan of reform had been drawn up by Daunou, one of the authors of the Constitution of the Year III, who, in the Convention, had had Girondist sympathies. He did not wish to alter substantially the Frimaire laws, but he wanted to see, in addition, the establishment of 'complementary courses in medicine' in twenty-three provincial hospitals [39]; there doctors would be able to improve their knowledge, and it would then be possible for local authorities to require proper qualifications for the practice of medicine:

> You will not re-establish guild-masterships, but you will require proof of capacity; one may become a doctor without having attended a school, but you will demand a solemn guarantee of the knowledge of every candidate: in this way, you will reconcile the rights of individual liberty with those of public safety [40].

There, even more clearly than before, the clinic appears as the concrete solution to the problem of the training of doctors and of the definition of medical competence.

Because of its timidity in reform and because of its fidelity to the principles of Year III, Daunou's project was unanimously criticized: Baraillon called it 'a prescription for organized murder' [41]. A few weeks later, the Commission d'Instruction Publique presented another report, drawn up this time by Calès. This second report was written in a quite different spirit: in order to win acceptance for a reconstitution of a professional body of doctors, which was implicit in his project, he opposed the distinction whereby physicians were confined to the towns, surgeons being 'all that was needed in the country', and apothecaries being entrusted with the treatment of children [42]. In the five schools to be set up in Paris, Montpellier, Nancy, Brussels, and Angers, physicians, surgeons, and apothecaries must attend the same courses. Studies would be checked by six examinations, which the students would take when they thought fit (a surgeon would need to take only three). Lastly, a jury, composed of doctors and pharmacists, would be set up in each department that 'would be consulted on all matters relating to the art of healing and to public health' [43]. Under the pretext of a more rational system of teaching, to be provided by a greater number of Faculties for all those concerned with public health, Calès's project aimed principally to re-establish a body of doctors qualified by a system of standardized studies and examinations.

Calès's project, supported by doctors like Baraillon and Vitet, was in turn violently attacked, from the outside by the École de Montpellier, which declared that it was satisfied with the measures taken by the Convention, and within the Assembly itself by all those who remained faithful to the spirit of Year III. Things dragged on. Taking advantage of the thwarting of the counter-revolution by the 18 Fructidor, Prieur de la Côte-d'Or, a former member of the Comité de Salut Public, succeeded in having Calès's project sent to the Commission d'Instruction Publique. He criticized it for the insignificant place it accorded to the clinic, and for its advocacy of a return to the teaching of the old Faculties: for 'it is not enough that the student should listen and read, he must also see, touch, and above all practise, and acquire the habit of practice' [44]. In this way, Prieur obtained a double tactical advantage: he showed the validity, at the scientific level, of the experience acquired by those who had more or less taught themselves medicine since 1792, and, stressing how expensive such clinical teaching was, he suggested that instead of increasing the number of schools, and thus sacrificing quality to quantity, only the school in Paris should be maintained. This amounted quite simply to a return to Fourcroy's project in its original form.

But meanwhile, on the very day before the uprising that was to reveal him as one of the leaders of the Royalist plot and so force him into exile, Pastoret had got a law passed through the Cinq-Cents concerned with the practice of medicine. A jury was to be set up for each of the three Écoles de Santé composed of two physicians, two surgeons, and a pharmacist whose task would be to supervise all those who wished to practise on their own; moreover, 'all those who are now practising the art of healing without having been legally received according to the forms laid down by ancient laws will be obliged to present themselves within three months' [45]. All those who had taken up medicine during the previous five years were therefore subjected to examination by juries trained in the old school; doctors would once again be able to control their own recruitment; they would be reconstituted as a body capable of defining their own criteria of competence.

The principle had gained acceptance, but the small number of Écoles de Santé made its application difficult; by demanding that they be reduced still further, Prieur thought that he would make the application of Pastoret's law impossible. In any case, this law remained a dead letter, and hardly four months had elapsed since it had been passed when the Directoire was compelled once again to draw the legislators' attention to the dangers that an uncontrolled medicine presented for citizens:

> A positive law should compel anyone who claims to practise one of the professions of the art of healing to undergo long studies and examination by a severe jury; science and custom must be respected, but incompetence and imprudence must be contained; public penalties should deter cupidity and suppress crimes that are little short of murder [46].

On 17 Ventôse Year VI, Vitet revived, before the Cinq-Cents, the main lines of Calès's project: five schools of medicine; in each department a council of health that would be concerned with epidemics 'and means of preserving the health of the inhabitants, and which would take part in the election of the professors; a series of four examinations to be held on fixed dates'. The only real innovation was the requirement of a clinical test: 'The candidate doctor will expound at the patient's bedside the character of the species of disease and its treatment.' Thus, for the first time, the criteria of theoretical knowledge and those of a practice that can be linked only to experience and custom were found together in a single institutional framework. Vitet's project did not permit the integration or gradual assimilation into official medicine of the 'free' medicine that had been practised since 1792; but it recognized theoretically, and in the framework of normal studies, the value of practice acquired in the hospitals. It was not 'free' medicine that was being recognized, but the value of experience as such in medicine.

Calès's plan had seemed too rigorous in the Year V; Vitet's plan, supported in turn by Calès and Baraillon, aroused as much opposition. It seemed quite clear that no reform of medical teaching would be possible until the problem for which it acted as a screen had been solved, namely, the problem of the practice of medicine. Calès's project having been rejected, Baraillon proposed to the

Cinq-Cents a resolution expressing in clear terms what had been its implicit meaning: no one could practise the art of healing unless he possessed qualifications deriving from either the new Schools or the old Faculties [47]. Porcher defended the same thesis in the Conseil des Anciens [48]. The whole problem was caught up in a political and conceptual impasse; but at least all these discussions had had the merit of revealing what the real question was: not the number or the programme of the Écoles de Santé, but the very meaning of the medical profession and the privileged character of the experience that it defines.

III. CABANIS'S INTERVENTION AND THE REORGANIZATION OF YEAR XI

Chronologically speaking, Cabanis presented his report on medical administration between Baraillon's project and the discussion of Vendémiaire in the Anciens, on 4 Messidor Year VI. In fact, this text already belonged to another age; it marked the stage at which ideology was to take an active, and often determining part in political and social restructuring. In this respect, Cabanis's text on medical administration is closer in spirit to the reforms of the Consulat than to the polemics contemporary with it. Although it attempted to define the conditions for a practical solution, it sought, above all, to provide the outline of a theory of the medical profession.

At the immediate, practical level, Cabanis dealt with two problems: that of the officers of health and that of examinations.

The senior officers presented no difficulties—they could be allowed to practise without further formalities. The others, however, would have to undergo an examination specially intended for them; it would be confined to 'the fundamental skills of the art, particularly those relating to its practice'. Ordinary medical studies, however, would have to be controlled by an examination, including a written test, an oral test, and 'exercises in anatomy, surgery, and internal and external clinical medicine'. Once the criteria of competence had been laid down, a selection could be made of those to whom the lives of citizens might be safely entrusted; medicine would then become a closed profession: 'Any person practising medicine who has not passed the examinations of the schools, or who has not appeared before the special juries, will be fined or, if the offence is repeated, committed to prison' [49].

The essential part of the text concerns the nature of the medical profession. The problem was to assign to it a closed domain, reserved to it alone, without either resorting to the corporative structures of the Ancien Régime or returning to forms of state control that might be reminiscent of the Convention period.

Taking industry in the wide sense of the term, Cabanis distinguishes between two categories of objects. There are objects whose nature is such that the consumers are themselves the judges of their utility: that is, public consciousness is sufficient to determine their value; this value is placed upon it by public opinion, is external to the object itself; it can have no secret, no error, no mystification, since it resides in a consensus. The idea of determining a value by decree had no more meaning than wishing to impose a truth upon it from the outside; real value can only be a free value:

> In a well-regulated social state, the freedom of industry must meet with no obstacle; it must be complete, unlimited; and as the development of an industry can become useful to him who cultivates it only insofar as it is useful to the public, it follows that the general interest is here truly at one with the particular interest.

But there are also industries whose object and value do not depend upon a collective decision: either these objects are among those that serve to determine the market value of other objects (precious metals, for example), or they relate to the human individual, about whom any error may prove fatal. Thus the value of an industrial object cannot be determined by consensus when it is itself a market criterion, or when it concerns, by its very existence, a member of the consensus. In either case, the industrial object has an intrinsic value that is not immediately visible: it is therefore subject to error and fraud; it must therefore be gauged. But how can the competent public be given an instrument of measurement that would itself involve competence? The public must delegate to the state control not over each of the objects produced (which would be contrary to the principles of economic freedom), but over the producer himself: the state must verify his capacity, his moral value, and, from time to time, 'the real value and quality of the objects that he produces'.

Therefore doctors should be supervised in the same way that goldsmiths are supervised, as men of secondary industry who do not produce wealth, but who treat that which measures or produces wealth: 'That is why physicians, surgeons, and pharmacists must be subject to stringent examination as to their knowledge, their abilities, and their moral habits. . . . This does not mean that industry will be impeded or the liberty of the individual infringed' [50].

Cabanis's proposition was not accepted; yet, in broad outline, it indicated the solution that was to be adopted, giving medicine the status of a liberal and protected profession that it has preserved up to the twentieth century. The law of 19 Ventôse Year XI concerning the practice of medicine conforms with Cabanis's themes and, in a more general way, with those of the Idéologues. It provided for a two-tier hierarchy in the medical body: doctors in medicine and surgery who had qualified in one of the six schools, and the officers of health, who would institutionalize in definitive form those whom Cabanis had wished to reintegrate on a provisional footing. After four examinations (anatomy and physiology; pathology and nosography; materia medica; hygiene and forensic medicine), doctors would take a test in clinical medicine, internal or external, according to whether they wished to become physicians or surgeons. The officers of health, who would provide 'the most ordinary care', would study for only three years in the schools, though even this would not be indispensable if they could prove that they had practised for five years in civil or military hospitals, or for six years as a doctor's private pupil or assistant. They would be examined by a department jury. Anyone not belonging to either of these two categories who dabbled in medicine would incur penalties ranging from a fine to imprisonment.

This whole movement of ideas, projects and measures between the Year VI and the Year XI had certain decisive significations.

1. In defining the closed character of the medical profession, one managed to avoid both the old corporative model and that control over medical acts themselves which was so repugnant to economic liberalism. The principle of choice and its control were based on the notion of competence, that is, on a set of possibilities that characterized the very person of the doctor: knowledge, experience, and that 'recognized probity' referred to by Cabanis [51]. The medical act is worth what he who has performed it is worth; his intrinsic value is a function of the socially recognized quality of the producer. Thus, within an economic liberalism patently inspired by Adam Smith, is defined a profession that is both 'liberal' and closed.

2. In this world of aptitudes, however, a difference of level was

introduced: on the one hand there were the 'doctors', and on the other the 'officers of health'. The old difference between physicians and surgeons, between the internal and the external, what one knows and what one sees, is made secondary by this new distinction. It is no longer a question of a difference in the object, or the way in which the object is manifested, but of a difference of level in the experience of the knowing subject. Between physicians and surgeons, there was already no doubt a hierarchy that was reflected in institutions: but it derived from an earlier difference in the objective domain of their activity; it was now displaced towards the qualitative index of this activity.

3. This distinction had an objective correlative: the officers of health would treat 'the industrious and active people' [52]. In the eighteenth century, it was accepted that the labouring classes, especially those in the country, led a more simple, moral, and healthy life than others, and were subject primarily to the external illnesses that came within the competence of the surgeon. From the Year XI, the distinction became a social one: one did not have to be 'learned and profound in theory' in order to treat the people, who often suffered from 'primitive accidents' and 'simple indispositions'; the officer of health would be quite experienced enough to deal with such matters. 'The history of the art, as that of men, shows that the nature of things, like the order of civilized societies, absolutely requires this distinction' [53]. In conformity with the ideal order of economic liberalism, the pyramid of qualities corresponded with the superposition of social strata.

4. On what was the distinction based among those practising the art of healing? The most important part of the training of an officer of health was his years of *practice*, which might be as many as six; the doctor, on the other hand, complemented his theoretical training with *clinical* experience. It was no doubt this difference between the practical and the clinical that was the most innovatory factor in the legislation of the Year XI. The practice required of the officer of health was a *controlled empiricism*: a question of knowing what to do after seeing; experience was integrated at the level of perception, memory, and repetition, that is, at the level of the example. In the clinic, it was a question of a much more subtle and complex structure in which the integration of experience occurred in a gaze that was at the same time knowledge, a gaze that exists, that was master of its truth, and free of all example, even if at times

it had made use of them. Practice would be *opened* up to the officers of health, but the doctors would *reserve* the initiation into the clinic to themselves.

This new definition of the clinic was bound up with a reorganization of the hospitals.

At first, both Thermidor and the Directoire reverted to the liberal principles of the Legislature; on 11 Thermidor Year III, Delecloy attacked the law providing for the nationalization of hospital funds on the ground that it placed the burden of medical care on the state alone, instead of placing it 'under the protection of general commiseration and under the guardianship of the rich' [54]. Between Pluviôse and Germinal Year IV, the government sent out to local administrations a series of circulars which, broadly speaking, reverted to the moral and economic criticisms that had been levelled, at the outset of the Revolution and even before, at the whole principle of hospitalization (the increased cost of an illness treated in a hospital, the lazy habits it induces, the financial distress and moral penury of a family deprived of a father or mother); it was hoped that there would be an increase in home treatment [55]. However, the time was past when such treatment was regarded as universally valid and when people dreamt of a society without alms-houses and hospitals: poverty was too widespread—there were over 60,000 paupers in Paris in the Year II [56] and their number was increasing; popular movements were too feared, and too much suspicion surrounded the political use to which individual assistance might be put, to allow the whole system of assistance to be left to them. A structure had to be found, for the preservation of both the hospitals and the privileges of medicine, that was compatible with the principles of liberalism and the need for social protection —the latter understood somewhat ambiguously as the protection of the poor by the rich and the protection of the rich against the poor.

One of the last acts of the Thermidorian Convention was to suspend, on 2 Brumaire Year IV, the execution of the law to nationalize hospital funds. On the basis of a new report submitted by Delecloy on 12 Vendémiaire Year IV, the law of 23 Messidor was definitively revoked: the funds that had been sold would be replaced by national funds, and the government would thereby be discharged of all obligation. The hospitals would recover their civil character; their organization and management were entrusted to the municipal administrations, which would appoint a five-member executive committee. This municipalization of the hospitals freed the state from any necessity of providing assistance and left the burden of identifying themselves with the poor to fairly small-scale collectivities: each commune became responsible for its own poverty and for the way in which it protected itself from it. The system of obligation and compensation between rich and poor no longer passed through the law of the state, but, by means of a sort of contract, subject to variation in space and suspension in time, it belonged more to the order of free consent.

A stranger, more hidden contract of the same kind was silently being formed about the same time between the hospital, where the poor were treated, and the clinic, in which doctors were trained. Once again, the thinking of those last days of the Revolution revived, sometimes word for word, what had been formulated in the period immediately preceding it. The most important moral problem raised by the idea of the clinic was the following: by what right can one transform into an object of clinical observation a patient whose poverty has compelled him to seek assistance at the hospital? He had asked for help of which he was the absolute subject, insofar as it had been conceived specifically for him; he was now required to be the object of a gaze, indeed, a relative object, since what was being deciphered in him was seen as contributing to a better knowledge of others. Furthermore, while observing, the clinic was also carrying out research; and this search for the new exposed it to a certain amount of risk: a doctor in private practice, Aikin remarked [57], must take care of his reputation; his way must be that of safety, if not of certainty; 'In the hospital he is not fettered in this way and his genius may express itself in a new way.' Does not the very essence of hospital aid become altered by the following principle: 'Hospital patients are, for several reasons, the most suitable subjects for an experimental course'? [58]

A certain balance must be kept, of course, between the interests of knowledge and those of the patient; there must be no infringement of the natural rights of the sick, or of the rights that society owes to the poor. The domain of the hospital was an ambiguous one: theoretically free, and, because of the non-contractual character of the relation between doctor and patient, open to the indifference of experiment, it bristled with obligations and moral limitations deriving from the unspoken—but present—contract binding man in general to poverty in its universal form. If, in the hospital, the doctor does not carry out theoretical experiments, free of all obligation to their human object, it is because, as soon as he sets foot in the hospital, he undergoes a decisive moral experience that circumscribes his otherwise unlimited practice by a closed system of duty. 'It is by entering the asylums where poverty and sickness languish together that he will feel those painful emotions, that active commiseration, that burning desire to bring comfort and consolation, that intimate pleasure that springs from success, and which the sight of spreading happiness cannot but increase. It is there that he will learn to be religious, humane, compassionate' [59].

But to look in order to know, to show in order to teach, is not this a tacit form of violence, all the more abusive for its silence, upon a sick body that demands to be comforted, not displayed? Can pain be a spectacle? Not only can it be, but it must be, by virtue of a subtle right that resides in the fact that no one is alone, the poor man less so than others, since he can obtain assistance only through the mediation of the rich. Since disease can be cured only if others intervene with their knowledge, their resources, their pity, since a patient can be cured only in society, it is just that the illnesses of some should be transformed into the experience of others; and that pain should be enabled to manifest itself: 'The sick man does not cease to be a citizen. . . . The history of the illnesses to which he is reduced is necessary to his fellow men because it teaches them by what ills they are threatened.' If he refused to offer himself as an object of instruction, the patient would be guilty of ingratitude, because 'he would have enjoyed the advantages resulting from sociability, without paying the tribute of gratitude' [60]. And in accordance with a structure of reciprocity, there emerges for the rich man the utility of offering help to the hospitalized poor: by paying for them to be treated, he is, by the same token, making possible a greater knowledge of the illnesses with which he himself may be affected; what is benevolence towards the poor is transformed into knowledge that is applicable to the rich:

> Beneficent gifts will assuage the ills of the poor from which enlightenment will result for the preservation of the rich. Yes, rich benefactors, generous men, this sick man lying in the bed that

you have subscribed is now experiencing the disease that will be attacking you ere long; he will be cured or perish; but in either event, his fate may enlighten your physician and save your life [61].

These, then, were the terms of the contract by which rich and poor participated in the organization of clinical experience. In a regime of economic freedom, the hospital had found a way of interesting the rich; the clinic constitutes the progressive reversal of the other contractual part; it is the *interest* paid by the poor on the capital that the rich have consented to invest in the hospital; an interest that must be understood in its heavy surcharge, since it is a compensation that is of the order of *objective interest* for science and of *vital interest* for the rich. The hospital became viable for private initiative from the moment that sickness, which had come to seek a cure, was turned into a spectacle. Helping ended up by paying, thanks to the virtues of the clinical gaze.

These themes, which were so characteristic of pre-Revolutionary thinking, and which found frequent expression before the Revolution, were given new meaning and immediate application in the liberalism of the Directoire. Explaining in the Year VII how the maternity clinic at Copenhagen functioned, Demangeon asserted, against all objections of modesty and discretion, that only 'unmarried women, or those who claimed to be such' were admitted. 'It seems that nothing better could be imagined, for it is precisely that class of women whose feelings of modesty are likely to be the least delicate' [62]. Thus, this morally disarmed and socially so dangerous class may be of the greatest possible use to honourable families; morality will find its reward in that which flouts it, for the women 'not being in a state to exercise beneficence . . . at least contribute to the training of good doctors and repay their benefactors with interest' [63].

The doctor's gaze is a very small saving in the calculated exchanges of a liberal world. . . .

NOTES

[1] Vicq d'Azyr, *Oeuvres* (Paris, 1805, vol. V, p. 64).
[2] Demangeon, *Du moyen de perfectionner la médecine*, p. 29.
[3] Cantin, *Projet de réforme adressé à l'Assemblée* (Paris, 1790, p. 13).

[4] Lioult, *Les charlatans dévoilés* (Paris, Year VIII, preface, unpaged).
[5] A.N. 17, A 1146, d. 4, quoted by A. Seboul, *Les Sans-Culottes parisiens en l'an II* (Paris, 1958, p. 494, n. 127).
[6] Message from the Directoire to the Conseil des Cinq-Cents, 24 Nivôse Year VI, quoted by Baraillon in his report of 6 Germinal Year VI.
[7] 22 Brumaire and 4 Frimaire Year V.
[8] Message of 24 Nivôse Year VI.
[9] P. Rambaud, *L'Assistance publique à Poitiers jusqu'à l'an V*, Vol. II, p. 200.
[10] Guillaume, *Procès-verbaux du Comité d'Instruction publique de la Convention*, vol. IV, pp. 878-9.
[11] Baraillon, *Rapport au Conseil des Cinq-Cents*, 6 Germinal Year VI, p. 6, concerning the scandal of amputations.
[12] *Ibid.*
[13] *Opinion de Porcher au Conseil des Anciens*, session of 16 Vendémiaire Year VI, pp. 14-15.
[14] By the section of Lombards, cf. Soboul, *op. cit.*, p. 495.
[15] 'Adresse de la section de l'Homme armé, des Invalides et Lepeletier à la Convention', *ibid.*
[16] Hospital for pregnant women set up by the Section du Contrat social.
[17] E. Pastoret, *Rapport fait au nom de la Commission d'Instruction publique sur un mode provisoire d'examen pour les officiers de santé*, 16 Thermidor Year V, p. 2.
[18] A. Girbal, *Essai sur l'esprit de la clinique médicale de Montpellier* (Montpellier, 1858, pp. 7-11).
[19] Fourcroy, *Rapport et projet de décret sur l'enseigement libre des sciences et des arts*, Year II, p. 2.
[20] Fourcroy, *Rapport à la Convention au nom des Comités de Salut public et d'Instruction publique*, 7 Frimaire Year III, p. 3.
[21] *Ibid.*, p. 3.
[22] *Ibid.*, p. 6.
[23] *Ibid.*, p. 9.
[24] *Ibid.*, p. 10.
[25] *Ibid.*, pp. 12-13.
[26] *Plan général de l'enseignement dans l'École de Santé de Paris* (Paris, Year III, p. 11).
[27] *Ibid.*, p. 39.
[28] *Ibid.*, p. 1.
[29] *Ibid.*, pp. 1-2.
[30] Baraillon, *op. cit.*, p. 2.
[31] *Plan général de l'enseignement dans l'École de Santé de Paris*, Year III, p. 31.
[32] *Opinion de J.-Fr. Baraillon*, session of the Assemblée des Cinq-Cents, 17 Germinal Year VI, p. 4.
[33] Prospectus accompanying the first number of the *Recueil périodique de la Société de Santé de Paris*.
[34] *Ibid.*, I, p. 3.

[35] *Ibid.*, II, p. 234.
[36] *Mémoires de la Société médicale d'émulation*, vol. I, Year V, p. II.
[37] *Ibid.*, p. IV.
[38] From March 1798, Cabanis sat in the Assemblée des Cinq-Cents, as a representative of the Institut.
[39] P.-C.-F. Daunou, *Rapport à l'Assemblée des Cinq-Cents sur l'organisation des écoles spéciales*, 25 Floréal Year V, p. 26.
[40] *Ibid.*
[41] Baraillon, *Rapport au Conseil des Anciens*, 6 Germinal Year VI, p. 2.
[42] *Rapport de J.-M. Calès sur les Écoles spéciales de Santé*, 12 Prairial Year V, p. 11.
[43] *Ibid.*, Articles 43-6.
[44] *Motion d'ordre de C.A. Prieur relative au projet sur les Écoles de Santé*, Session of the Cinq-Cents, 12 Brumaire Year V, p. 4.
[45] *Rapport fait par Pastoret sur un mode provisoire d'examen pour les officiers de Santé*, 16 Thermidor Year V, p. 5.
[46] Message from the Directoire to the Assemblée des Cinq-Cents, 24 Nivôse Year VI.
[47] Baraillon, *Rapport à l'Assemblée des Cinq-Cents sur la partie de la police qui tient à la médecine*, 6 Germinal Year VI.
[48] Porcher, *Opinion sur le mode provisoire d'examen pour les Officiers de Santé* (Assemblée des Anciens), 16 Vendémiaire Year VI.
[49] Cabanis, *Rapport du conseil des cinq cents sur un mode provisoire de police médicale*, 4 Messidor Year VI, pp. 12-18.
[50] *Ibid*, pp. 6-7.
[51] *Ibid.*
[52] Quoted as a reference by J.-C.-F. Caron, *Réflexions sur l'exercise de médecine* (Paris, Year XII).
[53] Fourcroy, *Discours prononcé au corps Législatif le 19 Ventôse an XI*, p. 3.
[54] Quoted by Imbert, *Le droit hospitalier sous la Révolution et l'Empire*, p. 93, n. 94.
[55] *Ibid.*, p. 104, n. 3.
[56] Cf. Soboul, *Les Sans-Culottes parisiens en l'an II* (Paris, 1958).
[57] J. Aikin, *Observations sur les hôpitaux* (Fr. trans., Paris, 1777, p. 104).
[58] *Ibid.*, p. 103.
[59] Menuret, *Essai sur les moyens de former de bons médecins* (Paris, 1791, pp. 56-7).
[60] Chambon de Montaux, *Moyen de rendre les hôpitaux plus utiles à la nation* (Paris, 1787, pp. 171-2).
[61] Du Laurens, *Moyens de rendre les hôpitaux utiles et de perfectionner la médecine* (Paris, 1787, p. 12).
[62] J.-B. Demangeon, *Tableau historique d'un triple établissement réuni en un seul hospice à Copenhague* (Paris, Year VII, pp. 34-5).
[63] *Ibid.*, pp. 35-6.

6 · Signs and Cases

And here we have the unbounded extent of the clinical domain:

Unravel the principle and cause of an illness through the confusion and obscurity of the symptoms; know its nature, its forms, its complications; distinguish at first glance all its characteristics and differences; by means of a prompt and delicate analysis separate it from all that is foreign to it; foresee what beneficial or detrimental events might occur in the course of its duration; use the favourable moments that nature provides to effect a solution; calculate the forces of life and the activity of the organs; augment or diminish their energy as required; determine precisely when you should act and when it would be better to wait; decide confidently between several methods of treatment all of which offer advantages and inconveniences; choose the one whose effects seem most rapid, most agreeable, and most certain of success; benefit from experience; seize your opportunities; calculate your chances and your risks; make yourself master of your patients and their affections; assuage their pains; calm their anxieties; anticipate their needs; bear with their whims; make the most of their characters and command their will, not as a cruel tyrant reigns over his slaves, but as a kind father who watches over the destiny of his children [1].

This solemn, prolix text yields its meaning in the light of another statement, which, paradoxically, through its sheer brevity can be superimposed: 'One must, as far as possible, make science ocular' [2]. So many powers, from the slow illumination of obscurities, the ever-prudent reading of the essential, the calculation of

88

times and risks, to the mastery of the heart and the majestic confiscation of paternal authority, are just so many forms in which the
sovereignty of the gaze gradually establishes itself—the eye that
knows and decides, the eye that governs.

The clinic was probably the first attempt to order a science on
the exercise and decisions of the gaze. From the second half of the
seventeenth century, natural history had set out to analyse and
classify natural beings according to their visible characters. All this
'treasure' of knowledge that antiquity and the Middle Ages had
accumulated—and which concerned the virtues of plants, the
powers of animals, secret correspondences and sympathies—since
Ray, all this had become marginal knowledge for naturalists. What
remained to be discovered, however, were 'structures', that is,
forms, spatial arrangements, the number and size of elements: natural history took upon itself the task of mapping them, of tran-
scribing them in discourse, of preserving, confronting, and combining them, in order to make it possible, on the one hand, to
determine the vicinities and kinships of living beings (and therefore
the unity of creation) and, on the other, to recognize rapidly any
individual (and therefore his unique place in creation).

The clinic demands as much of the gaze as natural history. As
much, and to a certain extent, the same thing: to see, to isolate
features, to recognize those that are identical and those that are
different, to regroup them, to classify them by species or families.
The naturalist model, to which medicine had partly been subjected
in the eighteenth century, remained active. The old dream of
Boissier de Sauvages of being the Linnaeus of diseases was not entirely forgotten in the nineteenth century: doctors long continued
to botanize in the field of the pathological. But the medical gaze
was also organized in a new way. First, it was no longer the gaze
of any observer, but that of a doctor supported and justified by an
institution, that of a doctor endowed with the power of decision
and intervention. Moreover, it was a gaze that was not bound by
the narrow grid of structure (form, arrangement, number, size),
but that could and should grasp colours, variations, tiny anomalies,
always receptive to the deviant. Finally, it was a gaze that was not
content to observe what was self-evident; it must make it possible
to outline chances and risks; it was calculating.

It would be untrue, no doubt, to see in late eighteenth-century
clinical medicine a mere return to the purity of a gaze long bur-

dened with false knowledge. It is not even a question of a displacement of this gaze, or of a finer application of its extent. New
objects were to present themselves to the medical gaze in the sense
that, and at the same time as, the knowing subject reorganizes himself, changes himself, and begins to function in a new way. It was
not, therefore, the conception of disease that changed first and later
the way in which it was recognized; nor was it the signaletic system
that was changed first and then the theory; but together, and at a
deeper level, the relation between the disease and this gaze to which
it offers itself and which at the same time it constitutes. At this
level there was no distinction to be made between theory and experience, methods and results; one had to read the deep structures
of visibility in which field and gaze are bound together by *codes
of knowledge*; in this chapter, we shall study these codes in their
two major forms: the linguistic structure of the sign and the aleatory structure of the case.

In the medical tradition of the eighteenth century, the disease
was observed in terms of *symptoms* and *signs*. These were distinguished from one another as much by their semantic value as by
their morphology. The symptom—hence its uniquely privileged
position—is the form in which the disease is presented: of all that
is visible, it is closest to the essential; it is the first transcription of
the inaccessible nature of the disease. Cough, fever, pain in the side,
and difficulty in breathing are not pleurisy itself—the disease itself
is never exposed to the senses, but 'reveals itself only to reasoning'—but they form its 'essential symptom', since they make it
possible to designate a pathological state (in contradistinction to
health), a morbid essence (different, for example, from pneumonia), and an immediate cause (a discharge of serosity) [3].
The symptoms allow the invariable form of the disease—set back
somewhat, visible and invisible—to *show through*.

The sign announces: the prognostic sign, what will happen; the
anamnestic sign, what has happened; the diagnostic sign, what is
now taking place. Between it and the disease is a distance that it
cannot cross without accentuating it, for it often appears obliquely
and unexpectedly. It does not offer anything to knowledge; at most
it provides a basis for recognition—a recognition that gradually
gropes its way into the dimensions of the hidden: the pulse betrays
the invisible strength and rhythm of the circulation; or, again, the

sign discloses time, just as the blueing of the nails is an unfailing announcement of death, or the crises of the fourth day, in intestinal fevers, promise recovery. Through the invisible, the sign indicates that which is further away, below, later. It concerns the outcome, life and death, time, not that immobile truth, that given, hidden truth that the symptoms restore to their transparency as phenomena.

Thus, the eighteenth century transcribed the double reality, natural and dramatic, of disease, establishing the truth of a corpus of knowledge and the possibility of its application. A happy, calm structure, in which a balance was struck between the Nature-Death system, with visible forms taking root in the invisible, and the Time-Outcome system, which anticipated the invisible by means of a visible mapping out (*repérage*).

Both these systems existed for themselves; their difference is a fact of nature to which medical perception adapted itself, but which it did not constitute.

The formation of the clinical method was bound up with the emergence of the doctor's gaze into the field of signs and symptoms. The recognition of its constituent rights involved the effacement of their absolute distinction and the postulate that henceforth the signifier (sign and symptom) would be entirely transparent for the signified, which would appear, without concealment or residue, in its most pristine reality, and that the essence of the signified—the heart of the disease—would be entirely exhausted in the intelligible syntax of the signifier.

I. THE SYMPTOMS CONSTITUTE A PRIMARY STRATUM INDISSOCIABLY SIGNIFIER AND SIGNIFIED

There is no longer a pathological essence beyond the symptoms: everything in the disease is itself a phenomenon; in that respect, the symptoms play a simple role, primary in nature: 'Their collection forms what is known as the disease' [4]. They are nothing more than a truth wholly given to the gaze; their link and status do not refer to an essence, but indicate a natural totality that has only its principles of composition and its more or less regular forms of duration: 'A disease is a whole, because one can assign it its elements; it has an aim, because one can calculate its results; it is therefore a whole placed between the limits of invasion and termination' [5]. The symptom has therefore lost its role of sovereign indicator, being merely a phenomenon of the law of appearance; it is on the same level as nature.

Yet not entirely so: something, in the immediacy of the symptom, signifies the pathological, which distinguishes it from a phenomenon belonging purely and simply to organic life. 'By phenomenon I mean any notable change in the healthy or sick body; hence the division into those that belong to health and those that designate disease: the latter are easily confused with the symptoms or sensible appearance of the disease' [6]. By this simple opposition to the forms of health, the symptom abandons its passivity as a natural phenomenon and becomes a signifier of the disease, that is, of itself taken as a whole, since the disease is simply a collection of symptoms. There is a strange ambiguity here, since in its signifying function the symptom refers both to the relation between phenomena themselves—to what constitutes their totality and the form of their coexistence—and to the absolute difference that separates health from disease; it signifies, therefore, by tautology, the totality of what it is and, by its emergence, the exclusion of what it is not. In its existence as pure phenomenon, it is indissociably the only nature of the disease, and the disease constitutes its only nature as a specific phenomenon. When it acts as a signifier in relation to itself, it is therefore doubly signified: by itself and by the disease, which, by characterizing it, opposes it to non-pathological phenomena; but, when taken as a signified (by itself or by the disease), it can receive its meaning only from an earlier act that does not belong to its sphere: from an act that totalizes and isolates it, that is, from an act that has transformed it into a sign in advance.

This complexity in the structure of the symptom is to be found in all philosophy of the natural sign; clinical thought merely transposes, into the more laconic and often more confused vocabulary of practice, a conceptual configuration whose discursive form was available, in all latitude, to Condillac. In the general equilibrium of clinical thought, the symptom plays more or less the role of the language of action: like it, it is caught up in the general movement of nature; and its force of manifestation is as primitive, as naturally given as the 'instinct' that bears this initial form of language [7]; it is the disease in its manifest state, just as the language of action is the impression itself in the animation that prolongs it, maintains it, and turns it back into an external form, which is of the same nature as its internal truth. But it is conceptually impossible that this immediate language should take on meaning for another's gaze, without the intervention of an act originating in another place: an act of which Condillac availed himself, in advance, by conferring consciousness upon the two speechless subjects (*sujets sans parole*), imagined in their immediate motility [8]; and whose singular, sovereign nature he has hidden by inserting it into the communicative, simultaneous movements of instinct [9]. When he posits the language of action as the origin of speech, Condillac slips secretly into it, by depriving it of any concrete figure (syntax, words, and even sounds), the linguistic structure inherent in each of the acts of a speaking subject. This enabled him to extract from the language of action language as such, since he had already inserted the possibility of language into the language of action. The same thing applies in the clinic, where the relations between this language of action, which is the symptom, and the explicitly linguistic structure of the sign are concerned.

II. IT IS THE SOVEREIGNTY OF CONSCIOUSNESS THAT TRANSFORMS THE
SYMPTOM INTO A SIGN

Signs and symptoms are and say the same thing, the only difference being that the sign *says* the same thing that *is* precisely the symptom. In its material reality, the sign is identified with the symptom itself; the symptom is the indispensable morphological support of the sign. Hence 'no sign without a symptom' [10]. But what makes the sign a sign belongs not to the symptom, but to an activity that originates elsewhere. Thus 'every symptom is a sign' by right, 'but not every sign is a symptom' [11] in the sense that the totality of symptoms will never be able to exhaust the reality of the sign. How does this operation occur, which transforms the symptom into a signifying element, and which signifies the disease as precisely as the immediate truth of the symptom?

By an operation that makes visible to itself the totality of the field of experience at each of its stages, and dissipates all its opaque structures:

—an operation that totalizes by comparing organisms: tumour, redness, heat, pain, throbbing, tension as a sign of phlegmon, because one hand is compared with another, one individual with another [12];

—an operation that recalls normal functioning: cold breath in

one subject is a sign of the disappearance of animal heat and, therefore, of a 'radical weakening of the vital forces and of their imminent destruction' [13];

—an operation that registers the frequency of simultaneity or
succession: 'What relation is there between a coated tongue, a
trembling of the lower lip, and a tendency to vomit? We do not
know, but observation has often shown the first two phenomena
accompanied by that state, and that is enough for them to become
signs in future' [14];

—lastly, an operation which, beyond first appearances, scruti-
nizes the body and discovers at the autopsy a visible invisible: thus
the examination of corpses has shown that in cases of pleuropneumonia with expectoration, the sudden interruption of pain and the
gradual weakening of the pulse beat are signs of a hepatization of
the lung.

Beneath a gaze that is sensitive to difference, simultaneity or
succession, and frequency, the symptom therefore becomes a sign—
a spontaneously differential operation, devoted to totality and to
memory, and calculating as well; an act, therefore, that joins, in a
single movement, the element and the connexion of the elements
among themselves. In that sense, it is really no more than Condillac's
analysis put into practice in medical perception. Here and there is
it not simply a question of 'composing and decomposing our ideas
in order to make different comparisons with them, and in order
to discover by this means the relations that they have among themselves, and the new ideas that they may produce'? [15] Analysis
and the clinical gaze also have this feature in common that they
compose and decompose only in order to reveal an ordering that is
the natural order itself: their artifice is to operate only in the resti-
tutive act of the original. 'This analysis is the true secret of discoveries because it makes us go back to the origin of things' [16].
For the clinic, this origin is the natural order of symptoms, the
form of their succession or of their reciprocal determination.
Between sign and symptom there is a decisive difference that
assumes value only against the background of an essential identity;
the sign is the symptom itself, but in its original truth. At last,
there emerges on the horizon of clinical experience the possibility
of an exhaustive, clear, and complete reading: for a doctor whose
skills would be carried 'to the highest degree of perfection, all
symptoms would become signs' [17], all pathological manifestations

would speak a clear, ordered language. One would at last be on a level with that serene, accomplished form of scientific knowledge, that 'well-made language' (*langue bien faite*) of which Condillac speaks.

III. THE BEING OF THE DISEASE CAN BE ENTIRELY STATED IN ITS TRUTH

Exernal signs taken from the pulse, heat, breathing, hearing, alteration in facial features, nervous or spasmodic affections, and impairment of the natural appetites form by their various combinations separate, more or less distinct, or strongly pronounced pictures. . . . A disease must be regarded as an indivisible whole from its beginning to its end, a regular set of characteristic symptoms and a succession of periods [18].

It is no longer a question of giving that *by which* the disease can be recognized, but of restoring, at the level of words, a history that covers its total being. To the exhaustive presence of the disease in its symptoms corresponds the unobstructed transparency of the pathological being with the syntax of a descriptive language: a fundamental isomorphism of the structure of the disease and of the verbal form that circumscribes it. The descriptive act is, by right, a 'seizure of being' (*une prise d'être*), and, inversely, being does not appear in symptomatic and therefore essential manifestations without offering itself to the mastery of a language that is the very speech of things. In the medicine of species, the nature of a disease and its description could not correspond without an intermediate stage that formed the 'picture' with its two dimensions; in clinical medicine, *to be seen* and *to be spoken* immediately communicate in the manifest truth of the disease of which it is precisely the whole *being*. There is disease only in the element of the visible and therefore statable.

The clinic brings into play what, for Condillac, was the fundamental relation between the perceptual act and the element of language. The clinician's description, like the philosopher's analysis, proffers what is given by the natural relation between the operation of consciousness and the sign. And in this repetition the order of natural connexions (*enchaînements*) is stated; far from perverting the logical necessities of time, the syntax of language restores them in their most original articulation: 'To analyse is simply to observe in a successive order the qualities of an object with a view to ascribing to them the simultaneous order in which they exist. . . . But what is this order? Nature herself indicates it; it is that in which it offers the object' [19]. The order of truth does only one thing with that of language, because both restore time to its necessary and statable, that is, *discursive* form. The *history* of diseases, to which Sauvages gave an obscurely spatial meaning, now assumes its chronological dimension. The *course* of time occupies in the structure of this new knowledge the role in classificatory medicine of the flat space of the nosological picture.

The opposition between nature and time, between what is manifested and what announces, has disappeared; the distinction between the essence of the disease, its symptoms and its signs, has also disappeared; and the play and distance by which the disease was manifested, but at a distance as it were, by which it betrayed itself, but at a distance and in uncertainty, have also disappeared. The disease escaped from this rotating structure of the visible that rendered it invisible and the invisible that rendered it visible, and dissipated itself in the visible multiplicity of symptoms that signified its meaning without remainder. The medical field was no longer to know these silent species, whether given or withdrawn; it was to open on to something which always speaks a language that is at one in its existence and its meaning with the gaze that deciphers it— a language inseparably read and reading.

As an isomorph of ideology, clinical experience offers it an immediate domain of application. Not that medicine, as Condillac supposed, had returned to an empirical respect for the thing perceived; but in the clinic, as in analysis, the armature of the real is designed on the model of language. The clinician's gaze and the philosopher's reflexion have similar powers, because they both presuppose a structure of identical objectivity, in which the totality of being is exhausted in manifestations that are its signifier-signified, in which the visible and the manifest come together in at least a virtual identity, in which the perceived and the perceptible may be wholly restored in a language whose rigorous form declares its origin. The doctor's discursive, reflective perception and the philosopher's discursive reflexion on perception come together in a figure of exact superposition, since *the world is for them the analogue of language*.

Medicine as an uncertain kind of knowledge is an old theme to which the eighteenth century was especially sensitive. It was to be found, reinforced by recent history, in the traditional opposition between the art of medicine and the knowledge of inert things: 'The science of man is concerned with too complicated an object, it embraces a multitude of too varied facts, it operates on too subtle and too numerous elements always to give to the immense combinations of which it is capable the uniformity, evidence, and certainty that characterize the physical sciences and mathematics' [20]. An uncertainty that was a sign of complexity concerning the object and of imperfection concerning science: no objective foundation was given to the conjectural character of medicine outside the relation between that extreme scantiness and that excessive richness.

Out of this defect the eighteenth century, in its last years, made a positive element of knowledge. In the period of Laplace, either under his influence or within a similar movement of thought, medicine discovered that uncertainty may be treated, analytically, as the sum of a certain number of isolatable degrees of certainty that were capable of rigorous calculation. Thus, this confused, negative concept, whose meaning derived from a traditional opposition to mathematical knowledge, was to be capable of transforming itself into a positive concept and offered to the penetration of a technique proper to calculation.

This conceptual transformation was decisive: it opened up to investigation a domain in which each fact, observed, isolated, then compared with a set of facts, could take its place in a whole series of events whose convergence or divergence were in principle measurable. It saw each perceived element as a *recorded event* and the uncertain evolution in which it found itself an *aleatory series*. It gave to the clinical field a new structure in which the individual in question was not so much a sick person as the endlessly reproducible pathological fact to be found in all patients suffering in a similar way; in which the plurality of observations was no longer simply a contradiction or confirmation, but a progressive, theoretically endless convergence; in which time was not an unforeseen element that might conceal, and which must be dominated by anticipatory knowledge, but a dimension to be integrated, since it introduces the elements of the series into its own course as so many degrees of certainty. Through the introduction of probabilistic thought, medicine entirely renewed the *perceptual values* of its domain: the space in which the doctor's attention had to operate became an unlimited space, made up of isolatable events whose form of solidarity was of the order of the series. The simple dialectic of the pathological species and the sick individual, an enclosed space and an uncertain time, was, in principle, dislocated. Medicine no longer tried to see the essential truth beneath the sensible individuality; it was faced by the task of perceiving, and to infinity, the events of an open domain. This was the clinic.

But at this period this schema was neither radicalized, thought out, nor even drawn up in an absolutely coherent way. It was not so much an over-all structure as a set of structural themes, juxtaposed with one another without having found their basis. Whereas in the case of the preceding configuration (sign-language) the coherence was real, though often only half-visible, here probability was being constantly invoked as a form of explanation or justification, but it only achieved a low degree of coherence. The reason for this did not lie in the mathematical theory of probabilities, but in the conditions that could make it applicable: the enumeration of physiological or pathological facts like that of a population or a series of astronomical events was not technically possible at a time when the hospital field was still so much on the fringe of medical experience that it often seemed to act as its caricature or distorting mirror. A conceptual mastery of probability in medicine implied the validation of a hospital domain which, in turn, could be recognized as a space of experience only by already probabilistic thinking. Hence the imperfect, precarious, and partial character of the calculation of certainties, and the fact that it sought for itself a confused basis that was opposed to its intrinsic technological meaning. Thus Cabanis tried to justify the instruments of the clinic, which were then still in the process of being formed, with the aid of a concept whose technical and theoretical level belonged to a much earlier accretion. He put aside the old concept of uncertainty only in order to reactivate the hardly better adapted one of the imprecise, free profusion of nature. This profusion 'brings nothing in exact precision: it seems to have wished to preserve a certain latitude for itself, in order to leave upon the movements that it imprints that regular liberty that never allows them to depart from order, but which renders them more varied and gives them more grace' [21]. But the important, conclusive part of the text is to be found in the note accompanying it: 'This latitude corresponds exactly to that which the art may possess in practice, or, rather, it provides it with its measure.' The imprecision that Cabanis attributes to the movements of nature is merely a void to be occupied by the technical armature of a perception of *cases*. The principal stages in this process are as follows:

1. COMPLEXITY OF COMBINATION.   The nosography of the eighteenth century implied a configuration of experience such that, however confused and complicated phenomena in their concrete presentation may be, they were related, more or less directly, to essences whose increasing generality guaranteed a decreasing complexity: the class was simpler than the species, which, in turn, was simpler than the actual, immediate disease, with all its phenomena and its modifications in a given individual. At the end of the eighteenth century, and in a demarcation of experience similar to Condillac's, simplicity is not to be found in the essential generality but at the primary level of the given, in the small number of endlessly repeated elements. It is not the class of fevers which, owing to the inadequate understanding of its concept, is the principle of intelligibility, but the small number of elements that are vital in a fever in every concrete case. The combinative variety of the simple forms constitutes empirical diversity:

> With each new case, one might think that we were presented with new facts; but they are merely different combinations, different subtleties: in the pathological state, there is never more than a small number of principal facts; all the others result from their combination and from their different degrees of intensity. The order in which they appear, their importance, their various relationships are enough to give birth to every variety of disease [22].

Consequently, the complexity of individual cases could no longer be attributed to those uncontrollable modifications that disturb essential truths, and force us to decipher them only in an act of recognition that neglects and abstracts; it may be grasped and recognized in itself, in a complete fidelity to everything it presents, if one analyses it according to the principles of a combination, that is, if one defines all the elements that compose it and the form of that composition. To know will therefore restore the movement by which nature associates. And it is in this sense that knowledge of life and life itself obey the same laws of genesis—whereas in classificatory thinking this coincidence could exist only once and in divine understanding, the progress of knowledge now had the same origin and found itself caught up in the same empirical process of becoming (*devenir*) as the progression of life: 'Nature wanted the source of our knowledge to be the same as that of life; we must receive impressions in order to live; we must receive impressions in order to know' [23]; and, in each case, the law of development is the law of combination of these elements.

2. THE PRINCIPLE OF ANALOGY.   The combinative study of elements revealed analogous forms of co-existence or succession that made it possible to identify symptoms and diseases. The medicine of species and classes also made use of them in the decipherment of pathological phenomena: the resemblance of disorders could be recognized from one case to another, just as the appearance of their reproductive organs could be recognized from one plant to another. But these analogies related only to inert morphological data: it was a question of perceived forms whose general lines could be superimposed, 'an inactive, constant state of bodies, a state foreign to the present nature of the function' [24]. The analogies on which the clinical gaze rested in order to recognize, in different patients, signs and symptoms are of a different order; they 'consist in the relations that exist first between the constituent parts of a single disease, and then between a known disease and a disease to be known' [25]. Thus understood, analogy is no longer a more or less close kinship that vanishes as one moves away from the essential identity; it is an isomorphism of relations between elements: it concerns a system of relations and reciprocal actions, a functioning or a dysfunctioning. Thus difficulty in breathing is a phenomenon that is found in much the same morphology in phthisis, asthma, heart disease, pleurisy, and scurvy: but it would be misleading and dangerous to attach too much importance to such a resemblance. The fruitful analogy that identifies a symptom is in relation to other functions or other disorders: muscular weakness (which is found in dropsy), a livid complexion (similar to that found in obstructions), spots on the skin (as in smallpox), and swollen gums (as that caused by an accumulation of tartar), form a constellation in which the co-existence of elements designates a functional interaction peculiar to scurvy [26]. It is the *analogy* of these relations that makes it possible to *identify* a disease in a series of diseases.

Furthermore, within the same disease and in the same patient, the principal of analogy may make it possible to define the singularity of the disease as a whole. Relying on the concept of sympathy, eighteenth-century doctors had used and abused the notion of 'complication', which always enabled them to find a pathological essence by simply extracting from the manifest symptoms whatever elements contradicted the essential truth, and these elements were then labelled as interferences. Thus gastric fever (fever, headaches, thirst, pain in the pit of the stomach) still conformed with its essence when accompanied by prostration, involuntary evacuations, a low and intermittent pulse rate, difficulty in swallowing: it was then described as being 'complicated' by an adynamic fever [27]. A rigorous use of analogy was to make it possible to avoid such arbitrariness in distinctions and groupings. From one symptom to another, in the same pathological entity, a certain analogy could be found in their relations with 'the external or internal causes that produced them' [28]. Thus many nosographers saw bilious pleuropneumonia as a complicated disease: if one saw the homology of relations existing between 'gastricity' (involving digestive symptoms and pains in the pit of the stomach) and irritation of the pulmonary organs, which suggests inflammation and respiratory disorders, different symptomatological sectors, apparently deriving from distinct morbid essences, make it possible nonetheless to give the disease its identity: that of a *complex figure* in the coherence of a unity, and not that of a *mixed reality* made up of mixed essences.

3. PERCEPTION OF FREQUENCIES. Medical knowledge will gain in certainty only in relation to the number of cases examined: this certainty 'will be complete if one extracts it from a mass of sufficient probability'; but if there is no 'rigorous deduction' of a sufficient number of cases, knowledge 'will remain of the order of conjecture and probability; it is no more than the simple expression of particular observations' [29]. Medical certainty is based not on the *completely observed individuality* but on the *completely scanned multiplicity of individual facts*.

By its multiplicity, the series becomes the vehicle of an index of convergence. Sauvages placed haemoptysis (the spitting of blood) among the haemorrhages and phthisis among the fevers—a distinction in accord with the structure of the phenomena that no symptomatic conjunction could challenge. But if the phthisis-haemoptysis complex (despite many distinctions according to individual cases, circumstances, stages) achieves a certain qualitative density in the total series, their connexion will become, over and above any encounter or any gap, outside even the manifest appearance of the phenomena, an essential relation: 'It is by studying the most frequent phenomena and meditating upon the order of their relations and their regular succession that one finds the bases of the general laws of nature' [30].

Individual variations are spontaneously effaced by integration. In the medicine of species, this effacement of particular modifications was assured only by a positive operation: in order to accede to the purity of essence, it was first necessary to possess it, and then to use it to obliterate the excessively rich content of experience; it was necessary, by a prior choice, 'to distinguish what is constant from what is variable in it, the essential from the purely accidental' [31]. In clinical experience, variations are not set aside, they separate of their own accord; they cancel each other out in the general configuration, because they are integrated into the domain of probability; they never fall outside the boundaries, however 'unexpected' or 'extraordinary' they may be; the abnormal is still a form of regularity: 'The study of monsters or of the monstrosities of the human species gives us an idea of nature's teeming resources and of the gaps to which she can lend herself' [32].

So we must abandon the idea of an ideal, transcendent Spectator whose genius and patience might be approached to a greater or lesser degree by real observers. The only normative observer is the totality of observers: the errors produced by their individual points of view are distributed in a totality that possesses its own powers of indication. Their very divergences reveal, in this nucleus in which, after all, they intersect, the outline of undeniable identities: 'Several observers never see the same fact in an identical way, unless nature has really presented it to them in the same way.'

Notions circulate, in obscurity and in an approximate vocabulary, in which one can recognize the calculation of error, the gap, the boundaries, the value of the average. All these notions indicate that the visibility of the medical field assumes a statistical structure and that medicine takes as its perceptual field not a garden of species but a domain of events. But nothing has become formalized as yet. And, curiously enough, it is in the effort to conceive a calculation of medical probabilities that failure, and the reasons for the failure, were to emerge.

In principle, this failure was due not to ignorance, or to a too superficial use of mathematical tools [33], but to the organization of the field.

4. THE CALCULATION OF THE DEGREES OF CERTAINTY.   'If one day one discovers in the calculation of probability a method that might be suitably adapted to complicated objects, to abstract ideas, to variable elements in medicine and physiology, one would soon produce the highest degree of certainty to which the sciences can attain [34]. It is a question of a calculus, which, from the outset, is valid within the domain of ideas, being both the principle of their analysis into constituent elements and a method of induction from frequences; it is offered, in an ambiguous way, as a logical and arithmetical distortion of approximation. The problem is, in fact, that late-eighteenth-century medicine never knew whether it was concerned with a series of facts whose laws of appearance and convergence were to be determined simply by the study of repetitions, or whether it was concerned with a set of signs, symptoms, and manifestations whose coherence was to be sought in a natural structure. It never ceased to hesitate between a *pathology of phenomena* and a *pathology of cases*. That is why the calculation of degrees of probability was immediately confused with the analysis of symptomatic elements: in a very strange way, it was the sign, as an element in a constellation, that was attributed, as a sort of natural right, with a coefficient of probability. But what had given it its value as a sign was not an arithmetic of cases but its connexion with a set of phenomena. Under the appearances of mathematics, the stability of a figure was gauged. The term 'degree of certainty' to be found in the writings of mathematicians designated, by a kind of crude mathematics, the more or less necessary character of an implication.

A simple example will enable us to grasp the nature of this fundamental confusion. Brulley recalls the principle formulated by Jacques Bernoulli in his *Ars conjectandi* that all certainty may be 'regarded as a whole divisible into as many probabilities as one wishes' [35]. Thus the certainty of pregnancy in a woman may be divided into eight degrees: the disappearance of menstruation; nausea and vomiting in the first month; an increase in the size of the womb in the second month; a much greater increase of the womb in the third month; the extension of the womb over the pubic bones; the projection of the whole hypogastric region in the fifth month; and the spontaneous movement of the foetus, which kicks against the internal surface of the womb; lastly, at the beginning of the last month, the movements of tossing and displacement [36]. Each of the signs, therefore, carries within itself one eighth of certainty: the succession of the first four constitutes a half-certainty, 'which forms the doubt itself, and may be regarded as a kind of balance'; beyond that probability begins [37]. This arithmetic of implication is valid for both curative indications and for diagnostic signs. A patient who had consulted Brulley wanted to be operated upon for a stone; there were two 'favourable probabilities' in favour of intervention: the good condition of the bladder and the small size of the stone; but there were four unfavourable probabilities against them: 'the patient is in his sixties; he is of the male sex; he is of a bilious temperament; he has a skin disease'. The subject would not hear of this simple arithmetic, and did not survive the operation.

It was hoped by an arithmetic of cases to balance the relation of logical structure; but between the phenomenon and what it signified there was not the same link as between the event and the series to which it belonged. This confusion occurred only because of the ambiguous virtues of analysis to which doctors were always having recourse: 'without the emblematic thread of analysis, we could not often find our way through the labyrinthine ways to the sanctuary of truth' [38]. But analysis was defined according to the *epistemological model* of mathematics and the *instrumental structure* of ideology. As an instrument, it served to define the system of implications in its complex totality: 'By this method, one dissects a subject, a complete idea; one examines the parts separately one after another, the most essential ones first, then those that are less so, with their various relations; one rises to the most simple idea'; but like its mathematical model, this analysis was used to determine an unknown idea: 'One examines the mode of composition, the way in which it has been operated, and hence, by the use of induction, one arrives at the unknown from the known' [39].

Selle said of the clinic that it was scarcely more than 'the very practice of medicine at the patient's bedside', and that, as such, it was identical with 'practical medicine in the strict sense' [40]. The clinic was much more than a revival of the old medical empiricism; it was concrete life, the first application of analysis. Moreover, despite its opposition to systems and theories, it recognizes its immediate kinship with philosophy: 'Why separate the science of the doctors from that of the philosophers? Why distinguish between two studies that share a common origin and end?' [41] The clinic is a field made philosophically 'visible' by the introduction into the pathological domain of grammatical and probabilistic structures. These structures may be dated historically, because they are contemporary with Condillac and his successors. They freed medical perception from the play of essence and symptoms, and from the no less ambiguous play of species and individuals: the figure disappeared by which visible and invisible were pivoted in accordance with the principle that the patient both conceals and reveals the specificity of his disease. A domain of clear visibility was opened up to the gaze.

But are not this domain itself and that which, fundamentally, makes it visible doubly in accord? Are they not based on overlapping forms that nevertheless evade one another? The grammatical model, acclimatized in the analysis of signs, remains implicit and enveloped without formalization in the depths of the conceptual movement: it is a question of a *transference of the forms of intelligibility*. The mathematical model is always explicit and invoked; it is present as the principle of coherence of a conceptual process that culminates outside itself: it is a question of the *contribution of themes of formalization*. But this fundamental contradiction was not felt to be such. And the gaze that rested on this apparently liberated domain seemed, for a time, a happy gaze.

NOTES

[1] C.-L. Dumas, *Éloge de Henri Fouquet* (Montpellier, 1807), quoted by A. Girbal, *Essai sur l'esprit de la clinique médicale de Montpellier* (Montpellier, 1858, p. 18).
[2] M.-A. Petit, *Discours sur la manière d'exercer la bienfaisance dans les hôpitaux*, 3 November 1797, in *Essai sur la médecine du coeur*, p. 103.
[3] Cf. Zimmermann, *Traité de l'expérience* (Fr. trans., Paris, 1774, vol. I, pp. 197–8).
[4] J.-L.-V. Broussonnet, *Tableau élémentaire de la séméiotique* (Montpellier, Year VI, p. 60).
[5] Audibert-Caille, *Mémoire sur l'utilité de l'analogie en médecine* (Montpellier, 1814, p. 42).
[6] J.-L.-V. Broussonnet, *op. cit.*, p. 59.

[7] Condillac, 'Essai sur l'origine des connaissances humaines', *Oeuvres complètes, Year VI*, vol. I, p. 262.

[8] *Ibid.*, p. 260.

[9] *Ibid.*, pp. 262-3.

[10] A.-J. Landré-Beauvais, *Séméiotique* (Paris, 1813, p. 4).

[11] *Ibid.*

[12] Favart, *Essai sur l'entendement médical* (Paris, 1822, pp. 8-9).

[13] Landré-Beauvais, *op. cit.*, p. 5.

[14] *Ibid.*, p. 6.

[15] Condillac, *op. cit.*, p. 109.

[16] *Ibid.*

[17] Demorcy-Delettre, *Essai sur l'analyse appliquée au perfectionnement de la médecine* (Paris, 1810, p. 102).

[18] Ph. Pinel, *La médecine clinique* (3rd edn., Paris, 1815, introduction, p. vii).

[19] Condillac, quoted by Pinel, *Nosographie philosophique* (Paris, Year VI, introduction, p. xi).

[20] C.-L. Dumas, *Discours sur les progrès futurs de la science de l'homme* (Montpellier, Year XII, pp. 27-8).

[21] Cabanis, *Du degré de certitude de la médecine* (3rd edn., Paris, 1819, p. 125).

[22] *Ibid.*, pp. 86-7.

[23] *Ibid.*, pp. 76-7.

[24] Audibert-Caille, *op. cit.*, p. 13.

[25] *Ibid.*, p. 30.

[26] C.-A. Brulley, *De l'art de conjecturer en médecine* (Paris, 1801, pp. 85-7).

[27] Pinel, *op. cit.*, p. 78.

[28] Audibert-Caille, *op. cit.*, p. 31.

[29] Dumas, *op. cit.*, p. 28.

[30] F.-J. Double, *Séméiologie générale* (Paris, 1811, vol. I, p. 33).

[31] Zimmermann, *op. cit.*, vol. I, p. 146.

[32] Double, *op. cit.*, vol. I, p. 33.

[33] Brulley, for example, was well acquainted with the writings of Bernoulli, Condorcet, and S'Gravesandy. Cf. Brulley, *op. cit.*, pp. 33-57.

[34] Dumas, *op. cit.*, p. 29.

[35] Brulley, *op. cit.*, pp. 26-7.

[36] *Ibid.*, pp. 27-30.

[37] *Ibid.*, pp. 31-2.

[38] Roucher-Deratte, *Leçons sur l'art d'observer* (Paris, 1807, p. 53).

[39] *Ibid.*

[40] Selle, *Introduction à l'étude de la nature* (Fr. trans., Paris, Year III, p. 229).

[41] Dumas, *op. cit.*, p. 21.

7 · Seeing and Knowing

'Hippocrates applied himself only to observation and despised all systems. It is only by following in his footsteps that medicine can be perfected' [1]. But the privileges that the clinic had recently recognized in observation were much more numerous than the prestige accorded it by tradition and of a quite different nature. They were at the same time the privileges of a pure gaze, prior to all intervention and faithful to the immediate, which it took up without modifying it, and those of a gaze equipped with a whole logical armature, which exorcised from the outset the naïvety of an unprepared empiricism. We must now describe the concrete exercise of such a perception.

The observing gaze refrains from intervening: it is silent and gestureless. Observation leaves things as they are; there is nothing hidden to it in what is given. The correlative of observation is never the invisible, but always the immediately visible, once one has removed the obstacles erected to reason by theories and to the senses by the imagination. In the clinician's catalogue, the purity of the gaze is bound up with a certain silence that enables him to listen. The prolix discourses of systems must be interrupted: 'All theory is always silent or vanishes at the patient's bedside' [2]; and the suggestions of the imagination—which anticipate what one perceives, find illusory relations, and give voice to what is in-accessible to the senses—must also be reduced: 'How rare is the accomplished observer who knows how to await, in the silence of the imagination, in the calm of the mind, and before forming his judgement, the relation of a sense actually being exercised!' [3] The gaze will be fulfilled in its own truth and will have access to the truth of things if it rests on them in silence, if everything keeps silent around what it sees. The clinical gaze has the paradoxical ability to *hear a language* as soon as it *perceives a spectacle*. In the clinic, what is manifested is originally what is spoken. The opposition between clinic and experiment overlays exactly the difference between the language we hear, and consequently recognize, and the question we pose or, rather, impose: 'The observer . . . reads nature, he who experiments questions' [4]. To this extent, observation and experiment are opposed but not mutually exclusive: it is natural that observation should lead to experiment, provided that experiment should question only in the vocabulary and within the language proposed to it by the things observed; its questions can be well founded only if they are answers to an answer itself without question, an absolute answer that implies no prior language, because, strictly speaking, it is the first word. It is this privilege of possessing an unsupersedable (*indépassable*) origin that the Double expresses in terms of causality: 'observation must not be confused with experience; the latter is the result or effect, the former the means or cause; observation leads naturally to experience' [5]. The observing gaze manifests its virtues only in a double silence: the relative silence of theories, imaginings, and whatever serves as an obstacle to the sensible immediate; and the absolute silence of all language that is anterior to that of the visible. Above the density of this double silence things seen can be heard at last, and heard solely by virtue of the fact that they are seen.

This gaze, then, which refrains from all possible intervention, and from all experimental decision, and which does not modify, shows that its reserve is bound up with the strength of its armature. To be what it must be, it is not enough for it to exercise prudence or scepticism; the immediate on which it opens states the truth only if it is at the same time its origin, that is, its starting point, its principle and law of composition; and the gaze must restore as truth what was produced in accordance with a genesis: in other words, it must reproduce in its own operations what has been given in the very movement of composition. It is precisely in this sense that it is 'analytic'. Observation is logic at the level of perceptual contents; and the art of observing seems to be

> a logic for those meanings which, more particularly, teach their operations and usages. In a word, it is the art of being in relation with relevant circumstances, of receiving impressions from objects

as they are offered to us, and of deriving inductions from them that are their correct consequences. Logic is . . . the basis of the art of observing, but this art might be regarded as one of the parts of Logic whose object is more dependent on meanings [6].

One can, therefore, as an initial approximation, define this clinical gaze as a perceptual act sustained by a logic of operations; it is analytic because it restores the genesis of composition; but it is pure of all intervention insofar as this genesis is only the syntax of the language spoken by things themselves in an original silence. The gaze of observation and the things it perceives communicate through the same Logos, which, in the latter, is a genesis of totalities and, in the former, a logic of operations.

Clinical observation involves two necessarily united domains: the hospital domain and the teaching domain.

The hospital domain is that in which the pathological fact appears in its singularity as an event and in the series surrounding it. Not long ago the family still formed the natural locus in which truth resided unaltered. Now its double power of illusion has been discovered: there is a risk that disease may be masked by treatment, by a regime, by various actions tending to disturb it; and it is caught up in the singularity of physical conditions that make it incomparable with others. As soon as medical knowledge is defined in terms of frequency, one no longer needs a natural environment; what one now needs is a neutral domain, one that is homogeneous in all its parts and in which comparison is possible and open to any form of pathological event, with no principle of selection or exclusion. In such a domain everything must be possible, and possible in the same way.

What a source of instruction is provided by two infirmaries of 100 to 150 patients each! . . . What a varied spectacle of fevers or phlegmasias, malign or benign, sometimes highly developed in strong constitutions, sometimes in a slight, almost latent, condition, together with all the forms and modifications that age, mode of life, seasons, and more or less energetic moral affections can offer! [7]

The old objection that the hospital causes modifications that are both pathological disorders and disorderings of pathological forms is neither dismissed nor ignored but rigourously annulled, since the modifications in question are uniformly valid for all events; it is possible, therefore, to isolate them by analysis and to treat them separately; by setting aside modifications due to locality, season, and nature of treatment 'one can succeed in introducing into the hospital clinic and general medical practice a degree of foresight and precision' [8]. The clinic is not, therefore, that mythical landscape in which diseases appear of their own accord, completely revealed; it makes possible the integration, in experience, of the hospital modification in a constant form. What the medicine of species called *nature* is shown to be merely the discontinuity of heterogeneous and artificial conditions; the 'artificial' diseases of the hospital permit pathological events to be reduced to the homogeneous; the hospital domain is no doubt not pure transparency to truth, but the refraction that is proper to it makes possible, through its constancy, the analysis of truth.

By means of the endless play of modifications and repetitions, the hospital clinic makes possible, therefore, the setting aside of the extrinsic. But this same play makes possible the summation of the essential in knowledge: in fact, variations cancel each other out, and the effect of the repetition of constant phenomena outlines spontaneously the fundamental conjunctions. By showing itself in a repetitive form, the truth indicates the way by which it may be acquired. It offers itself to knowledge by offering itself to recognition. 'The student . . . cannot familiarize himself overmuch with the repeated sight of alterations of all kinds, whose particular practice might later show him the picture' [9]. The genesis of the manifestation of truth is also the genesis of the knowledge of truth. There is, therefore, no difference in nature between the clinic as science and the clinic as teaching. A group is thus formed consisting of the master and his pupils, in which the act of recognition and the effort to know find fulfillment in a single movement. In its structure and in its two aspects as manifestation and acquisition, medical experience now has a collective subject; it is no longer divided between those who know and those who do not; it is made up, as one entity, of those who unmask and those before whom one unmasks. The statement is the same; the disease speaks the same language to both.

The *collective* structure of medical experience, the *collective* character of the hospital field—the clinic is situated at the meeting point of the two totalities; the experience that defines it traverses the surface of their confrontation and of their reciprocal boundary. There it derives not only its inexhaustible richness but also its sufficient, enclosed form. It is the carving up of the infinite domain of events by the intersection of the gaze and mutual questions. At the Edinburgh clinic, observation consisted of four series of questions: the first concerned the patient's age, sex, temperament, and occupation; the second, his symptoms; the third, the origin and development of the disease; and the fourth, more distant causes and earlier accidents [10]. Another method—one used at Montpellier—consisted of a general examination of all the visible modifications of the organism: 'first, the alterations of the body in general; second, those in the matter excreted; third, those denoted by the exercise of the functions' [11]. Pinel levelled the same criticism at both forms of investigation: they were unlimited. To the first, he objected: 'How, in the midst of this profusion of questions . . . can one grasp the essential, specific features of the disease?' and to the second, in corresponding fashion: 'What an immense enumeration of symptoms . . . ! Will this not throw us back into a new chaos?' [12] The questions to be asked are innumerable; the things to be seen infinite. If the clinical domain is open only to the tasks of language or to the demands of the gaze, it will have no limits and, therefore, no organization. There is boundary, form, and meaning only if interrogation and examination are connected with each other, defining at the level of fundamental structures the 'meeting place' of doctor and patient. In its initial form, the clinic seeks to determine this locus by three means:

1. THE ALTERNATION OF SPOKEN STAGES AND PERCEIVED STAGES IN AN OBSERVATION. In the schema of the ideal investigation sketched by Pinel, the general indication of the first stage is visual: one observes the present state in its manifestations. But the questionnaire already guarantees the place of language within this examination; the symptoms that first strike the senses of the observer are noted, but immediately afterwards the patient is questioned as to the pains he feels, and lastly, by observation, the state of the most important physiological functions is described. The second stage is dominated by language as well as by time, memory, developments, and successive incidents. First what, at a given moment, was perceptible must be recognized (recalling the forms of invasion, the succession of symptoms, the appearance of their present characteristics, and

the medicaments already applied). Then the patient or his entourage
must be questioned as to his general appearance, his occupation, his
past life. The third stage of observation is again one of perception;
a day-by-day account is kept of the progress of the disease under
four headings: evolution of the symptoms, possible appearance of
new phenomena, state of the secretions, and effect of medicaments
used. The final stage is reserved to speech: the prescription of the
regime during convalescence [13]. In the event of death, most
clinicians—but, as we shall see, Pinel less readily than others—
reserved to the gaze the final, most decisive authority, namely, the
anatomy of the body. In this regular alternation of speech and gaze,
the disease gradually declares its truth, a truth that it offers to the
eye and ear, whose theme, although possessing only one *sense*, can
be restored, in its indubitable totality, only by two *senses*: that
which sees and that which listens. This is why the questionnaire
without the examination and the examination without the interrogation were doomed to an endless task: it belongs to neither to fill the
gaps within the province of the other.

2. THE EFFORT TO DEFINE A STATUTORY FORM OF CORRELATION
BETWEEN THE GAZE AND LANGUAGE.   The theoretical and practical
problem confronting the clinicians was to know whether it would
be possible to introduce into a spatially legible and conceptually
coherent representation that element in the disease that belongs to a
visible symptomatology and that which belongs to a verbal analysis.
This problem was revealed in a technical difficulty that was very
revealing of the demands of clinical thinking: the *picture*. Is it
possible to integrate into a picture, that is, into a structure that is
at the same time visible and legible, spatial and verbal, that which is
perceived on the surface of the body by the clinician's eye, and that
which is heard by that same clinician in the essential language of the
disease? Perhaps the most naïve attempt was made by Fordyce:
he included in the x axis all the notations concerning the climate,
the seasons, prevalent diseases, the patient's temperament, idiosyncrasy, appearance, age, and previous accidents; and he classified in
the y axis the symptoms according to the organ or function in which
they were manifested (pulse, skin, temperature, muscles, eyes,
tongue, mouth, breathing, stomach, intestines, urine) [14]. It is
clear that this functional distinction between visible and expressible
(*énonçable*), and their correlation in the myth of an analytic
geometry, could be of no use in the work of clinical thought; such

an effort is significant only of the data of the problem and of the terms to be correlated. The pictures drawn up by Pinel seem simpler: their conceptual structure is in fact more subtle. As in Fordyce, the y axis includes the symptomatic elements that the disease offers to perception; but in the x axis, he indicates the significant values that these symptoms may assume. In an acute fever, a painful sensitivity in the pit of the stomach, a headache, and a violent thirst are to be included in a gastric symptomatology; on the other hand, prostration and abdominal tension have an adynamic meaning; lastly, pain in the limbs, a dry tongue, rapid breathing, a paroxysm, especially one occurring in the evening, are signs of both gastricity and adynamism [15]. Thus each visible segment assumes a significant value, and the picture certainly serves an analytical function in clinical knowledge. But it is obvious that the analytical structure is neither produced nor revealed by the picture itself; the analytical structure preceded the picture, and the correlation between each symptom and its symptomatological value was fixed once and for all in an essential a priori; beneath its apparently analytical function, the picture's only role is to divide up the visible within an already given conceptual configuration. The task is not, therefore, one of correlation, but, purely and simply, of redistribution of what was given by a perceptible extent in a conceptual space defined in advance. It makes nothing known; at most, it makes possible recognition.

3. THE IDEAL OF AN EXHAUSTIVE DESCRIPTION.   The arbitrary or tautological appearance of these pictures led clinical thought towards another form of correlation between the visible and the expressible, namely, the continuous correlation of an entirely— that is, doubly—faithful description; in relation to its object it must be, in effect, without any gap; and in language describing the object it must allow no deviation. Descriptive *rigour* will be the result of *precision* in the statement and of *regularity* in the designation: which, according to Pinel, is 'the method now followed in all other parts of natural history' [16]. Thus language is charged with a dual function: by its value as precision, it establishes a correlation between each sector of the visible and an expressible element that corresponds to it as accurately as possible; but this expressible element operates, within its role as description, a denominating function which, by its articulation upon a constant, fixed vocabulary, authorizes comparison, generalization, and establishment within

a totality. By virtue of this dual function, the work of description
ensures 'a prudent reserve in rising to general views without lend-
ing reality to abstract terms' and 'a simple, regular distribution, invariably based on the relations of structure or the organic functions
of the parts' [17].

It is in this exhaustive and complete passage from the *totality of
the visible* to the *over-all structure of the expressible (structure
d'ensemble de l'énonçable)* that is fulfilled at last that significative
analysis of the perceived that the naïvely geometric architecture of
the picture failed to provide. It is description, or, rather, the implicit
labour of language in description, that authorizes the transformation
of symptom into sign and the passage from patient to disease and
from the individual to the conceptual. And it is there that is forged,
by the spontaneous virtues of description, the link between the
random field of pathological events and the pedagogical domain in
which they formulate the order of their truth. To describe is to
follow the ordering of the manifestations, but it is also to follow the
intelligible sequence of their genesis; it is to see and to know at the
same time, because by saying what one sees, one integrates it
spontaneously into knowledge; it is also to learn to see, because it
means giving the key of a language that masters the visible. The
well-made language, which Condillac and his successors saw as the
ideal of scientific knowledge, must not therefore be sought, as do
certain over-hasty doctors [18], on the side of a language of
calculation, but on the side of that *measured language* that has the
measure of both the things that it describes and the language in
which it describes them. For the dream of an arithmetical structure
of medical language must be substituted, therefore, the search for a
certain internal measurement consisting of fidelity and fixity, of
primary and absolute openness to things and rigour in the considered
use of semantic values. 'The art of describing facts is the supreme
art in medicine: everything pales before it' [19].

Over all these endeavours on the part of clinical thought to
define its methods and scientific norms hovers the great myth of a
pure Gaze that would be pure Language: a speaking eye. It would
scan the entire hospital field, taking in and gathering together each
of the singular events that occurred within it; and as it saw, as it
saw ever more and more clearly, it would be turned into speech
that states and teaches; the truth, which events, in their repetitions
and convergence, would outline under its gaze, would, by this same
gaze and in the same order, be reserved, in the form of teaching, to

those who do not know and have not yet seen. This speaking eye would be the servant of things and the master of truth.

It is understandable that, after the revolutionary dream of an absolutely open science and practice, a certain medical esotericism could be revived around these themes: one now sees the visible only because one knows the language; things are offered to him who has penetrated the closed world of words; and if these words communicate with things, it is because they obey a rule that is intrinsic to their grammar. This new esotericism is different in structure, meaning, and use from that which made Molière's doctors speak in Latin: then it was simply a matter of not being understood and of preserving at the level of linguistic formulation the corporate privileges of a profession; now operational mastery over things is sought by accurate syntactic usage and a difficult semantic familiarity with language. Description, in clinical medicine, does not mean placing the hidden or the invisible within reach of those who have no direct access to them; what it means is to give speech to that which everyone sees without seeing—a speech that can be understood only by those initiated into true speech. 'Whatever precepts are given about so delicate a matter, it will always remain beyond the reach of the multitude' [20]. Here, at the level of theoretical structures, we encounter once again the theme of initiation, the outline of which is already to be found in the institutional forms of the same period [21]: we are at the heart of the clinical experience —a form of the *manifestation* of things in their truth, a form of *initiation* into the truth of things. It was this that Bouillaud was to declare as a self-evident banality some forty years later: 'The medical clinic may be regarded either as a science or as a way of teaching medicine' [22].

A hearing gaze and a speaking gaze: clinical experience represents a moment of balance between speech and spectacle. A precarious balance, for it rests on a formidable postulate: that all that is *visible* is *expressible*, and that it is *wholly visible* because it is *wholly expressible*. A postulate of such scope could permit a coherent science only if it was developed in a logic that was its rigorous outcome. But the reversibility, without residue, of the visible in the expressible remained in the clinic a requirement and a limit rather than an original principle. Total *description* is a present and ever-withdrawing horizon; it is much more the dream of a thought than a basic conceptual structure.

There is a simple historical reason for this: Condillac's logic did not allow a science in which the visible and the describable were caught up in a total adequation. Condillac's philosophy gradually shifted from an analysis of the original impression to an operational logic of signs, then from this logic to the constitution of a knowledge that would be both language and calculation: used at these three levels, and each time with different meanings, the notion of *element* sustained throughout this reflexion an ambiguous continuity, but one without a defined, coherent logical structure. Condillac never derived a universal logic from the element—whether this element was perceptual, linguistic, or calculable; he never ceased to hesitate between two logics of operations: of genesis and of calculation. Hence the dual definition of analysis: reduce complex ideas 'to the simple ideas of which they are made up and follow the progress of their generation' [23]; and seek the truth 'by a kind of calculation, that is, by composing and decomposing notions and comparing them in the most favourable way with the discoveries that one has in view' [24].

This ambiguity had its effect on clinical method, but this method followed a conceptual 'slope' that was the exact opposite of Condillac's development: the term by term reversal of the point of origin and the point of culmination.

It redescended from the exigency of calculation to the primacy of genesis; after seeking to define the postulate of equation of the visible with the expressible by a *universal*, rigorous calculability, it gave that postulate the meaning of total, exhaustive *description*. The essential operation was no longer combinative but a matter of syntactic transcription. Nothing is more typical of this movement —which takes up again, in the opposite direction, Condillac's whole approach—than Cabanis's thought, and this is particularly apparent if we compare it with Brulley's analysis. Brulley wished 'to regard certainty as a whole divisible into as many probabilities as one may wish'. 'A probability is therefore a degree, a part of certainty from which it differs as the part differs from the whole' [25]; medical certainty must thus be obtained by a combination of probabilities; after laying down the rules of such a combination Brulley declares that he will go no further, that he must leave to a more celebrated doctor the task of elucidating this subject—a task that he would have great difficulty in carrying out [26]. In all probability, it was Cabanis to whom he referred. For in *Les Révolutions de la médecine* the certain form of science is not defined by a type of calculation but by an organization whose values are essentially expressive; it is not a question of drawing up a calculation to proceed from the probable to the certain, but of determining a syntax in order to proceed from the element of the perceived to the coherence of discourse: 'the theoretical part of a science must, therefore, be the simple statement of the sequence of classification and of the relationship of all the facts which make up this science; it must, so to speak, be its summary expression' [27]. And if Cabanis finds room for the calculation of probabilities in the construction of medicine, it is only as one element among others in the total construction of scientific discourse. Brulley tried to place himself at the level of *La Langue des calculs*; although Cabanis cited this text, his thought is structurally on a footing with the *Essai sur l'origine des connaissances*.

It might be thought—and all the clinicians of that generation thought so—that things would rest there and that an unproblematic equilibrium was possible at that level between the composition of the visible and the syntactic rules of the expressible. But this was to be no more than a brief period of euphoria, a golden age with no future, in which seeing, saying, and learning to see by saying what one saw communicated in an immediate transparence: experience was rightfully science; and 'knowing' was in step with 'learning'. The gaze saw sovereignty in a world of language whose clear speech it gathered up effortlessly in order to restore it in a secondary, identical speech: given by the visible, this speech, without changing anything, made it possible to see. In its sovereign exercise, the gaze took up once again the structures of visibility that it had itself deposited in its field of perception.

But this generalized form of transparence leaves opaque the status of the language that must be its foundation, its justification, and its delicate instrument. Such a deficiency, which also occurs in Condillac's logic, opens up the field to a number of epistemological myths that are destined to mask it. But these myths are already engaging the clinic in new spatial figures, in which visibility thickens and becomes cloudy, in which the gaze is confronted by obscure masses, by impenetrable shapes, by the black stone of the body.

1. THE FIRST OF THESE EPISTEMOLOGICAL MYTHS CONCERNS THE ALPHABETICAL STRUCTURE OF DISEASE. At the end of the eighteenth

century, the alphabet appeared to grammarians to be the ideal schema of analysis and the ultimate form of the decomposition of a language; by that very fact it constituted the way in which that language was learnt. This alphabetical image was transposed essentially unaltered into the definition of the clinical gaze. The smallest possible observable segment, that from which one must set out and beyond which one cannot go back, is the singular impression one receives of a patient, or, rather, of a symptom of that patient; it signifies nothing in itself, but assumes meaning and value and begins to speak if it blends with other elements:

> Particular, isolated observations are to science what letters and words are to discourse; discourse is founded only on the concourse and coming together of letters and words whose mechanism and value must have been studied and reflected upon before correct and practical use was made of them; the same may be said of observations [28].

This alphabetical structure of disease ensures not only that one can always return to the 'unsupersedable' (*indépassable*) element; it also ensures that the number of these elements will be finite and even small. It is not first impressions that are diverse and apparently infinite, but their combination within a single disease: just as the small number of 'modifications designated by the grammarians under the name of consonants' is enough to give 'to the expression of feeling the precision of thought', so, for pathological phenomena, 'with each new case, one might think that one is presented with new facts, whereas they are merely new combinations of facts. In the pathological state, there is never more than a small number of principal phenomena. . . . The order in which they appear, their importance, and their various relations are enough to give birth to every variety of disease' [29].

2. THE CLINICAL GAZE EFFECTS A NOMINALIST REDUCTION ON THE ESSENCE OF THE DISEASE.   Composed as they are of letters, diseases have no other reality than the order of their composition. In the final analysis, their varieties refer to those few simple individuals, and whatever may be built up with them and above them is merely Name. And name in a double sense: in the sense in which the Nominalists use it when they criticize the substantial reality of abstract, general beings; and in another sense, one closer to a philoso-

phy of language, since the form of composition of the being of the disease is of a linguistic type. In relation to the individual, concrete being, disease is merely a name; in relation to the isolated elements of which it is made up, it has all the rigorous architecture of a verbal designation. To ask what is the essence of a disease is like 'asking what is the nature of the essence of a word' [30]. A man coughs; he spits blood; he has difficulty in breathing; his pulse is rapid and hard; his temperature is rising; these are all so many immediate impressions, so many letters, as it were. Together, they form a disease, pleurisy: 'But what, then, is pleurisy? . . . It is the concourse of the accidents that constitute it. The word pleurisy merely retraces them in a more abbreviated manner.' 'Pleurisy' has no more being than the word itself; it 'expresses an abstraction of the mind'; but, like the word, it is a well-defined structure, a multiple figure 'in which all or almost all the accidents are combined. If one or more are lacking, it is no longer pleurisy, or at least not real pleurisy' [31]. Disease, like the word, is deprived of being, but, like the word, it is endowed with a configuration. The nominalist reduction of existence frees a constant truth. That is why:

3. THE CLINICAL GAZE OPERATES ON PATHOLOGICAL PHENOMENA A REDUCTION OF A CHEMICAL TYPE.   Until the end of the eighteenth century the gaze of the nosographers was a gardener's gaze; one had to recognize the specific essence in the variety of appearances. At the beginning of the nineteenth century another model emerged: that of the chemical operation, which, by isolating the component elements, made it possible to define the composition, to establish common points, resemblances, and differences with other totalities, and thus to found a classification that was no longer based on specific types but on forms of relations: 'Instead of following the example of the botanists, should not the nosologists have, rather, taken as their model the systems of the chemist-mineralogists, that is, be content to classify the elements of diseases and their more frequent combinations?' [32] The notion of analysis in which, applied to the clinic, we have already recognized a quasi-linguistic sense and a quasi-mathematical sense [33] will now move towards a chemical signification: it will have as its horizon the isolation of pure bodies and the depiction of their combinations.. One has passed from the theme of the combinative to that of syntax and finally to that of combination.

And, by reciprocity, the clinician's gaze becomes the functional equivalent of fire in chemical combustion; it is through it that the essential purity of phenomena can emerge: it is the separating agent of truths. And just as combustions reveal their secret only in the very vividness of fire, and it would be useless to ask, once the flame was extinguished, what can remain in the inert powders, the *caput mortuum*, so it is in the act of voice and the brightness that it sheds over phenomena that truth is revealed: 'It is not the remains of the morbid combustion that the doctor should know, but the species of the combustion' [34]. The clinical gaze is a gaze that burns things to their furthest truth. The attention with which it observes and the movement by which it states are in the last resort taken up again in this paradoxical act of consuming. The reality, whose language it spontaneously reads in order to restore it as it is, is not as adequate to itself as might be supposed: its truth is given in a decomposition that is much more than a reading since it involves the freeing of an implicit structure. One can now see that the clinic no longer has simply to read the visible; it has to discover its secrets.

4. THE CLINICAL EXPERIENCE IS IDENTIFIED WITH A FINE SENSIBILITY.
    The clinical gaze is not that of an intellectual eye that is able to perceive the unalterable purity of essences beneath phenomena. It is a gaze of the concrete sensibility, a gaze that travels from body to body, and whose trajectory is situated in the space of sensible manifestation. For the clinic, all truth is sensible truth; 'theory falls silent or almost always vanishes at the patient's bedside to be replaced by observation and experience; for on what are observation and experience based if not on the relation of our senses? And where would they be without these faithful guides?' [35] And if this knowledge, at the level of the immediate use of the senses, is not attained at the outset, if it can acquire depth and mastery, it is not a shift in level that enables it to accede to something other than itself, it is a sovereignty that is entirely internal to its own domain; it only acquires depth at its own level, which is that of pure sensory perception; for sense can only spring from sense. What, then, is

> the doctor's glance, which so often involves such vast erudition and such solid instruction, if not the result of the frequent, methodical, and accurate exercise of the senses, from which derive that facility of application, that alertness to relations, that confidence of judge-

ment that is sometimes so rapid that all these acts seem to occur simultaneously, and are comprised together under the name of 'touch'? [36]

Thus this sensory knowledge—which nevertheless implies the conjunction of a hospital domain and a pedagogic domain, the definition of a field of probability and a linguistic structure of the real—is reduced to praise of the immediate sensibility.

The whole dimension of analysis is deployed only at the level of an aesthetic. But this aesthetic not only defines the original form of all truth, it also prescribes rules of exercise, and it becomes, at a secondary level, aesthetic in that it prescribes the norms of an art. The sensible *truth* is now open, not so much to the senses themselves, as to a *fine* sensibility. The whole complex structure of the clinic is summarized and fulfilled in the prestigious rapidity of an art: 'Since everything, or nearly everything, in medicine is dependent on a glance or a happy instinct, certainties are to be found in the sensations of the artist himself rather than in the principles of the art' [37]. The technical armature of the medical gaze is transformed into advice about prudence, taste, skill: what is required is 'great sagacity', 'great attention', 'great precision', 'great skill', 'great patience' [38].

At this level, all structures are dissolved, or, rather, those that constituted the essence of the clinical *gaze* are gradually, and in apparent disorder, replaced by those that are to constitute the *glance*. And they are very different. In fact, the gaze implies an open field, and its essential activity is of the successive order of reading; it records and totalizes; it gradually reconstitutes immanent organizations; it spreads out over a world that is already the world of language, and that is why it is spontaneously related to hearing and speech; it forms, as it were, the privileged articulation of two fundamental aspects of *saying* (what is said and what one says). The glance, on the other hand, does not scan a field: it strikes at one point, which is central or decisive; the gaze is endlessly modulated, the glance goes straight to its object. The glance chooses a line that instantly distinguishes the essential; it therefore goes beyond what it sees; it is not misled by the immediate forms of the sensible, for it knows how to traverse them; it is essentially demystifying. If it strikes in its violent rectitude, it is in order to shatter, to lift, to release appearance. It is not burdened with all the abuses of language. The glance is silent, like a finger pointing, denouncing.

There is no statement in this denunciation. The glance is of the non-verbal order of *contact*, a purely ideal contact perhaps, but in fact a more *striking* contact, since it traverses more easily, and goes further beneath things. The clinical eye discovers a kinship with a new sense that prescribes its norm and epistemological structure; this is no longer the ear straining to catch a language, but the index finger palpating the depths. Hence that metaphor of 'touch' (*le tact*) by which doctors will ceaselessly define their glance [39].

And by that very fact, clinical experience sees a new space opening up before it: the tangible space of the body, which at the same time is that opaque mass in which secrets, invisible lesions, and the very mystery of origins lie hidden. The medicine of symptoms will gradually recede, until it finally disappears before the medicine of organs, sites, causes, before a clinic wholly ordered in accordance with pathological anatomy. The age of Bichat has arrived.

NOTES

[1] Clifton, *État de la médecine ancienne et moderne* (Paris, 1742, preface by the translator, unpaged).
[2] Corvisart, Preface to the French translation of Auenbrugger, *Nouvelle méthode pour reconnaître les maladies internes de la poitrine* (Paris, 1808, p. vii).
[3] *Ibid.*, p. viii.
[4] Roucher-Deratte, *Leçons sur l'art d'observer* (Paris, 1807, p. 14).
[5] Double, *Séméiologie générale*, vol. I, p. 80.
[6] Sénebier, *Essai sur l'art d'observer et de faire des expériences* (2nd edn., Paris, 1802, Vol. I, p. 6).
[7] Ph. Pinel, *Médecine clinique* (Paris, 1815, introduction, p. ii).
[8] *Ibid.*, p. i.
[9] Maygrier, *Guide de l'étudiant en médecine* (Paris, 1818, pp. 94-5).
[10] Pinel, *op. cit.*, p. 4.
[11] *Ibid.*, p. 3.
[12] *Ibid.*, pp. 5 and 3.
[13] *Ibid.*, p. 57.
[14] Fordyce, *Essai d'un nouveau plan d'observations médicales* (Fr. trans., Paris, 1811).
[15] Pinel, *op. cit.*, p. 78.
[16] Pinel, *Nosographie philosophique*, introduction, p. iii.
[17] *Ibid.*, pp. iii-iv.
[18] Cf. above, Chapter 6.
[19] Amard, *Association intellectuelle* (Paris, 1821, vol. I, p. 64).
[20] *Ibid.*, p. 65.

[21] Cf. above, Chapter 5.
[22] Bouillaud, *Philosophie médicale* (Paris, 1831, p. 244).
[23] Condillac, *Origine des connaissances humaines*, p. 162.
[24] *Ibid.*, p. 110.
[25] C.-A. Brulley, *Essai sur l'art de conjecturer en médecine*, pp. 26–7.
[26] *Ibid.*
[27] Cabanis, *Coup d'oeil sur les Révolutions et la Réforme de la médecine* (Paris, 1804, p. 271).
[28] Double, *op. cit.*, p. 79.
[29] Cabanis, *Du degré de certitude* (3rd edn., Paris, 1819, p. 86).
[30] *Ibid.*, p. 66.
[31] *Ibid.*, p. 66.
[32] Demorcy-Delettre, *Essai sur l'analyse appliquée au perfectionnement de la médecine* (Paris, 1818, p. 135).
[33] Cf. above, Chapter 6.
[34] Amard, *op. cit.*, vol. II, p. 389.
[35] Corvisart, *op. cit.*, p. vii.
[36] *Ibid.*, p. x.
[37] Cabanis, *op. cit.*, p. 126.
[38] Roucher-Deratte, *op. cit.*, pp. 87–99.
[39] Corvisart, *op. cit.*

8 · Open Up a Few Corpses

At a very early stage historians linked the new medical spirit with the discovery of pathological anatomy, which seemed to define it in its essentials, to bear it and overlap it, to form both its most vital expression and its deepest reason; the methods of analysis, the clinical examination, even the reorganization of the schools and hospitals seemed to derive their significance from pathological anatomy.

> An entirely new period for medicine has just begun in France . . . ; analysis applied to the study of physiological phenomena, an enlightened taste for the writings of Antiquity, the union of medicine and surgery, and the organization of the clinical schools have brought about an astonishing revolution that is characterized by progress in pathological anatomy [1].

Pathological anatomy was given the curious privilege of bringing to knowledge, at its final stage, the first principles of its positivity.

Why this chronological inversion? Why did time deposit at the end of its course what was contained at the outset, already opening up and justifying the way? For a hundred and fifty years, the same explanation had been repeated: medicine could gain access to that which founded it scientifically only by circumventing, slowly and prudently, one major obstacle, the opposition of religion, morality, and stubborn prejudice to the opening up of corpses. Pathological anatomy had had no more than a shadowy existence, on the edge of prohibition, sustained only by that courage in the face of malediction peculiar to seekers after secret knowledge; dissection was

carried out only under cover of the shadowy twilight, in great
fear of the dead: 'at daybreak, or at the approach of night', Valsalva
slipped furtively into graveyards to study at leisure the progress of
life and destruction'; later, Morgagni could be seen 'digging up the
graves of the dead and plunging his scalpel into corpses taken from
their coffins' [2]. With the coming of the Enlightenment, death,
too, was entitled to the clear light of reason, and became for the
philosophical mind an object and source of knowledge: 'When
philosophy brought its torch into the midst of civilized peoples, it
was at last permitted to cast one's searching gaze upon the inanimate
remains of the human body, and these fragments, once the vile prey
of worms, became the fruitful source of the most useful truths' [3].
A fine transmutation of the corpse had taken place: gloomy respect
had condemned it to putrefaction, to the dark work of destruction; in the boldness of the gesture that violated only to reveal, to
bring to the light of day, the corpse became the brightest moment
in the figures of truth. Knowledge spins where once larva was
formed.

This reconstitution is historically false. Morgagni had no difficulty in the middle of the eighteenth century in carrying out his
autopsies; nor did Hunter, some years later; the conflicts recounted
by his biographer are of an anecdotal character and indicate no
opposition on principle [4]. From 1754 the Vienna clinic had had
a dissection room; so had the clinic that Tissot had organized at
Pavia; at the Hôtel-Dieu in Paris, Desault was quite free 'to demon-
strate on the body deprived of life the alterations that had rendered
art useless' [5]. One has only to recall Article 25 of the Décret de
Marly: 'Let us urge magistrates and directors of hospitals to provide
the professors with corpses and so enable them to carry out their
anatomy demonstrations and to teach the operations of surgery' [6].
So there was no shortage of corpses in the eighteenth century, no
need to rob graves or to perform anatomical black masses; one was
already in the full light of dissection. By means of an illusion widespread in the nineteenth century, and one to which Michelet gave
the dimensions of a myth, history painted the end of the Ancien
Régime in the colours of the last years of the Middle Ages, confusing the upheavals of the Renaissance with the struggles of the
Enlightenment.

In the history of medicine, this illusion has a precise meaning;
it functions as a retrospective justification: if the old beliefs had
for so long such prohibitive power, it was because doctors had to

feel, in the depths of their scientific appetite, the repressed need to open up corpses. There lies the point of error, and the silent reason why it was so constantly made: the day it was admitted that lesions explained symptoms, and that the clinic was founded on pathological anatomy, it became necessary to invoke a transfigured history, in which the opening up of corpses, at least in the name of scientific requirements, preceded a finally positive observation of patients; the need to know the dead must already have existed when the concern to understand the living appeared. So a dismal conjuration of dissection, an anatomical church militant and suffering, whose hidden spirit made the clinic possible before itself surfacing into the regular, authorized, diurnal practice of autopsy, was imagined out of nothing.

But chronology is not so pliable: Morgagni published his *De sedibus* in 1760, and by means of Bonet's *Sepulchretum*, took his place in the great line derived from Valsalva; Lieutaud wrote a summary of the book in 1767. The corpse was part of the medical field, and this was unchallenged by religion and morality. Yet forty years later, Bichat and his contemporaries felt that they were *rediscovering* pathological anatomy from beyond a shadowy zone. A period of latency separates Morgagni's text and Auenbrugger's discovery from Bichat's and Corvisart's use of them, forty years that witnessed the formation of the clinical method. It is there that the point of repression lies, not in the survival of old memories: the clinic, a neutral gaze directed upon manifestations, frequencies, and chronologies, concerned with linking up symptoms and grasping their language, was, by its structure, foreign to the investigation of mute, intemporal bodies; causes and locales did not interest it: it was interested in history, not geography. Anatomy and the clinic were not of the same mind: strange as it may seem to us now that anatomy and the clinic are inseparably linked, and seem to us always to have been, it was clinical thought that for forty years prevented medicine from hearing the lesson of Morgagni. The conflict was not between a young corpus of knowledge and old beliefs, but between two types of knowledge. Before pathological anatomy could be readmitted into the clinic, a mutual agreement had to be worked out: on the one hand, new geographical lines, and, on the other, a new way of reading time. In accordance with this litigious arrangement, the knowledge of the living, ambiguous disease could be aligned upon the white visibility of the dead.

But for Bichat, the re-reading of Morgagni did not involve a break with the clinical experience that had just been acquired. On the contrary, fidelity to the method of the clinicians and, even beyond that method, the anxiety, which he shared with Pinel, to provide a basis for a nosological classification, remained essential. Paradoxically, the return to the questions of the *De Sedibus* was made on the basis of a problem in the grouping of symptoms and in the ordering of diseases.

Like the *Sepulchretum* and many other seventeenth- and eighteenth-century treatises, Morgagni's letters specified diseases by means of a local separation of their symptoms or point of origin. Anatomical dispersal was the directing principle of nosological analysis: frenzy, like apoplexy, belonged to diseases of the head; asthma, pleuropneumonia, and haemoptysis formed related species in that they were all three localized in the chest. Morbid kinship rested on a principle of organic proximity: the space that defined it was local. First the medicine of classifications and then the clinic had detached pathological analysis from this regionalism and constituted for it a space at once more complex and more abstract, concerned with order, successions, coincidences, and isomorphisms.

The major discovery of the *Traité des membranes*, later systematized in the *Anatomie générale*, is a principle of deciphering corporal space that is at once intra-organic, inter-organic, and trans-organic. The anatomical element has ceased to define the fundamental form of spatialization and to command, by a relation of proximity, the ways of physiological or pathological communication; it is now no more than a secondary form of a primary space, which, by a process of winding round, superposition, and thickening, constitutes it. This fundamental space is entirely defined by the thinness of the tissue; the *Anatomie générale* enumerates twenty-one: cells, the nervous tissue of animal life, the nervous tissue of organic life, arteries, veins, the tissue of the exhaling vessels, that of the absorbents, bones, medullary tissue, cartiles, fibrous tissue, fibro-cartilaginous tissue, animal muscular tissue, muscles, mucous membrane, serous membrane, synovial membrane, glands, the derma, the epidermis, and hair. The membranes are tissular individualities which, despite their often extreme tenuity, 'are linked together only by indirect relations of organization with neighbouring parts' [7]. A general gaze often confuses them with the organ that they envelop and define; dissection of the heart has sometimes been carried out in which the pericardium was not distinguished, or of the lung without isolating the pleura; the peritoneum and the gastric organs were confused [8]. But a breakdown of these organic masses into tissular surfaces can and must be made if one is to understand the complexity of function and alteration: the hollow organs are provided with mucous membranes, covered with 'a fluid that usually moistens their free surface and that is supplied by small glands inherent in their structure'; the pericardium, the pleura, the peritoneum, and the arachnoid are serous membranes 'characterized by the lymphatic fluid that ceaselessly lubricates them and that is separated by exhalation from the volume of blood'; the periosteum, the dura mater, and the aponeuroses are made up of membranes 'that are moistened by no fluid' and that 'are composed of a white fibre similar to the tendons' [9].

On the basis of tissues alone, nature works with extremely simple materials. They are the elements of the organs, but they traverse them, relate them together, and constitute vast 'systems' above them in which the human body finds the concrete forms of its unity. There will be as many systems as there are tissues: in them, the complex, inexhaustible individuality of the organs is dispelled and suddenly simplified. Thus nature shows herself to be 'everywhere uniform in her procedures, variable only in their results, miserly of the means she employs, prodigal of the effects she obtains, modifying in a thousand different ways some few general principles' [10]. Between the tissues and the systems the organs appear as simple functional folds, entirely relative, both in their role and in their disorders, to the elements of which they are made up and to the groups to which they belong. Their density must be analysed and projected onto two surfaces: the particular surface of their membranes and the general surface of the systems. For the principle of diversification according to the organs that dominated the anatomy of Morgagni and his predecessors, Bichat substituted a principle of isomorphism in the tissues based on 'simultaneous identity and external conformation of structure, vital properties, and functions' [11].

Two very different structural perceptions were involved: Morgagni wished to perceive beneath the corporal surface the densities of the organs whose varied forms specified the disease; Bichat wished to reduce the organic volumes to great, homogeneous, tissual surfaces, to areas of identity in which secondary modification would find their fundamental kinships. In his *Traité des membranes*, Bichat imposes a diagonal reading of the body carried out according to expanses of anatomical resemblances that traverse the organs, envelop them, divide them, compose and decompose them, analyse them, and, at the same time, *bind them together*. It is the same form of perception as that borrowed by the clinic from Condillac's philosophy: the uncovering of an elementary that is also a universal, and a methodical reading that, scanning the forms of disintegration, describes the laws of composition. Bichat is strictly an analyst: the reduction of organic *volume* to tissular space is probably, of all the applications of analysis, the nearest to the mathematical model yet devised. Bichat's eye is a clinician's eye, because he gives an absolute epistemological privilege to the *surface gaze*.

The prestige that the *Traité des membranes* soon acquired is due, paradoxically, to that which separates it, essentially, from Morgagni, and places it in the line of clinical analysis: an analysis to which it brings, however, additional meaning.

Bichat's gaze is not a surface gaze in the sense in which clinical experience was a surface gaze. The tissual area is not an empty, imperceptible place where pathological events are offered to perception; it is a segment of perceptible space to which one can relate the phenomena of the disease. Thanks to Bichat, superficiality now becomes embodied in the real surfaces of membranes. Tissual expanses form the perceptual correlative of the surface gaze that defined the clinic. By a realistic shift in which medical positivism was to find its origin, surface, hitherto a structure of the onlooker, had become a figure of the one observed.

Hence the appearance that pathological anatomy assumed at the outset: that of an objective, real, and at last unquestionable foundation for the description of diseases: 'A nosography based on the affection of the organs will necessarily be invariable' [12]. In fact, tissual analysis makes it possible to draw up general pathological categories beyond Morgagni's geographical divisions; broad groups of diseases having the same major symptoms and the same type of evolution will emerge through organic space. All inflammations of serous membranes can be recognized by their thickening, the disappearance of their transparency, their whitish colour, their granulous alterations, and their adhesion to adjacent tissues. And just as the traditional nosologies began with a definition of the more general classes, pathological anatomy begins with 'a history of the alterations common to each system', whatever organ or region happens to be affected [13]. It would then be necessary to restore within each system the appearance assumed by the pathological phenomena according to the tissue. Inflammation takes the same form in all serous membranes but it does not attack all the tissues as easily or develop in them at the same speed: in decreasing order of susceptibility there is the pleura, the peritoneum, the pericardium, the vaginal canal, and, finally, the arachnoid [14]. The presence of tissues of the same texture throughout the organism makes it possible to see from one disease to another resemblances, kinships, and, in short, a whole system of communications inscribed in the deep configuration of the body. This non-local configuration is made up of interlocking concrete generalities, a whole organized system of implications. In fact, it really has the same logical armature as nosological thought. Beyond the clinic, which Bichat wishes to found and which is his starting point, he rediscovers not the geography of the organs, but the order of classifications. Pathological anatomy was *ordinal* before it was *localizing*.

Yet it gave to analysis a new, decisive value, showing, unlike the clinicians, that disease is the passive, confused object to which it must be applied only insofar as it is already, of itself, the active subject that exercises it pitilessly upon the organism. If the disease is to be analysed, it is because it is itself analysis; and ideological decomposition can be only the repetition in the doctor's consciousness of the decomposition raging in the patient's body. Although Van Horne, in the latter half of the seventeenth century, distinguished between arachnoid and pia mater, many authors, like Lieutaud, still confused the two. Alteration separates them clearly. During inflammation the pia mater reddens, showing that it is all vessel tissue; it then becomes harder and dryer. The diseased arachnoid becomes much whiter, and is covered with a viscous exudation; it alone can contract dropsy [15]. In the organic totality of the lung, pleurisy attacks only the pleura, pleuropneumonia the parenchyma, catarrhal coughs the mucous membranes [16]. Dupuytren showed that the effect of ligatures is not homogeneous throughout the whole thickness of the arterial duct: with pressure applied, the middle and internal walls cede and divide; only the cellulous, most external wall resists, because its structure is tighter

[17]. The principle of tissual homogeneity on which the general pathological types are based has as its correlative a principle of real division of the organs as a result of morbid alterations.

With his anatomy, Bichat does much more than provide the methods of analysis with a field of objective application; he makes analysis an essential stage in the pathological process. He realizes it within the disease, in the very web of its history. In a sense, nothing could be further removed from the implicit nominalism of the clinical method, in which analysis was directed, if not to words, at least to segments of perception that are always transcribable into language. One is now dealing with an analysis that is engaged in a series of real phenomena, and acting in such a way as to separate functional complexity into anatomical simplicities; it frees elements that are no less real and concrete for having been isolated by *abstraction*; in the heart, it reveals the pericardium, in the brain the arachnoid, in the intestines the mucous membranes. Anatomy could become pathological only insofar as the pathological spontaneously anatomizes. Disease is an autopsy in the darkness of the body, dissection alive.

This explains the enthusiasm that Bichat and his disciples immediately felt for the discovery of pathological anatomy: it was not that they rediscovered Morgagni beyond Pinel or Cabanis; they rediscovered analysis in the body itself; they revealed, in depth, the order of the surfaces of things; they defined for disease a system of *analytical classes* in which the element of pathological decomposition was the principle of generalization of morbid species. One passed from an analytical perception to the perception of real analyses. And, quite naturally, Bichat recognized in his discovery an event symmetrical with Lavoisier's: 'Chemistry has its simple bodies which form by the various combinations of which they are susceptible composite bodies. . . . Similarly, anatomy has its simple tissues which . . . by their combinations form organs' [18]. The method of the new anatomy is analysis, just as it is in chemistry, but an analysis detached from its linguistic support and defining the spatial divisibility of things rather than the verbal syntax of events and phenomena.

Hence the paradoxical reactivation of classificatory thought at the beginning of the nineteenth century. Pathological anatomy, which was to be proved right some years later, far from dissipating the old nosological project, gave it new vigour, insofar as it seemed to provide it with a solid basis: real analysis according to perceptible surfaces.

Astonishment has often been expressed that Bichat should have cited a text by Pinel concerning the principle of his discovery— Pinel, who until the end of his life was to remain deaf to the essential lessons of pathological anatomy. In the first edition of the *Nosographie*, Bichat read these sentences, which were like a revelation to him: 'What matter that the arachnoid, the pleura, and the peritoneum reside in different regions of the body, since these membranes have general conformities of structure? Are they not affected by similar lesions in the state of phlegmasia?' [19] This, in fact, was one of the first definitions of the principle of analogy applied to tissual pathology; but Bichat's debt to Pinel is still greater, since he found in the *Nosographie* the requirements, formulated though not satisfied, that this principle of isomorphism must fulfill: a classificatory analysis that makes possible a general ordering of the nosological picture. In the classification of diseases Bichat gave first place to 'the alterations common to each system', whatever the organ or region affected, but he accords this general form only to inflammations and scirrhi; other alterations are regional, and must be studied organ by organ [20]. Organic localization intervenes only as a residual method where the rule of tissual isomorphism cannot operate; Morgagni is used again only for lack of a more adequate reading of pathological phenomena. Laënnec considered that a better reading would become possible with time: 'It might be proved one day that almost all the different kinds of lesion may exist in all parts of the human body and that in each of these parts they present only slight modifications' [21]. Perhaps Bichat himself did not have enough confidence in his discovery, which, after all, was destined 'to change the face of pathological anatomy'; Laënnec believed that he had exaggerated the importance of the geography of the organs, to which one needed only to refer in order to analyse disorders of form and position (dislocations, hernias) and nutritional disorders (atrophies, hypertrophies); perhaps one day one might regard as belonging to the same pathological family the hypertrophies of the heart and those of the brain. On the other hand, Laënnec analyses, without regional boundaries, foreign bodies and especially alterations of texture, which have the same typology in all the tissual groupings: they are always either solutions of continuity (sores, fractures), accumulations or extravasations of natural liquids (fatty tumours, apoplexy), inflammations like pneumonia or gastritis, or accidental developments of tissues that did not exist before the disease. This is so in the case of scirrhi and tubercles [22]. At the time of Laënnec, Alibert tried to draw up a medical nomenclature modeled on chemistry: words ending in *osis* would designate the general forms of alteration (gastroses, leucoses, enteroses), those in *itis* would designate irritations of the tissues, those in *rhoea*, discharges, etc. And in concentrating solely on this project of fixing a meticulous, analytical vocabulary, he confuses (not flagrantly, because it was still conceptually possible) the themes of a nosology of a botanical type, those of localization in the manner of Morgagni, those of clinical description, and those of pathological anatomy:

> I use the method of the botanists already proposed by Sauvages . . . a method that consists in bringing together objects that have affinity with one another and in depositing those that have no similarity. In order to arrive at this philosophical classification, in order to give it fixed and invariable bases, I have grouped the diseases according to the organs that are their special sites. It will be seen that this was the only way of finding the characters that have most value for clinical medicine [23].

But how is it possible to adjust anatomical perception to the reading of symptoms? How could a simultaneous set of spatial phenomena establish the coherence of a temporal series that is, by definition, entirely anterior to it? From Sauvages to Double, the very idea of an atomical basis for pathology had had its adversaries, all convinced that the visible lesions on corpses could not designate the essence of an invisible disease. How in a complex lesional grouping can one distinguish the *essential order* from the *series of effects*? Are the lung adhesions in the body of a pleurisy patient one of the phenomena of the disease itself or a mechanical consequence of irritation? [24] There was the same difficulty in delineating the *original* and the *derived*: in a scirrhus of the pylorus one finds scirrhous elements in the epiploon and the mesentery; where should one place the first pathological fact? Lastly, anatomical signs are not a very good indicator of the intensity of the morbid process: there are very strong organic alterations that lead only to slight disturbances in the economy; but one would not suppose that a minuscule tumour on the brain could lead to death [25]. By never relating anything other than the visible, and in the simple, final, abstract form of its spatial coexistence, anatomy cannot say that which is connexion, process, and legible text in the order of time. A clinic of symptoms seeks the living body of the disease; anatomy provides it only with the corpse.

A doubly misleading corpse, too, since to the phenomena inter-rupted by death are added those caused by it and deposited on the organs in accordance with its own time scale. There are, of course, the phenomena of decomposition, which are difficult to dissociate from those belonging to the clinical picture of gangrene or putrid fever; on the other hand, there are phenomena of recession or effacement: the redness caused by irritations disappears very quickly after the cessation of the circulation; this interruption of natural movements (heartbeats, discharge of the lymph, breathing) itself causes effects whose beginning cannot be easily identified with that of the morbid elements: are the engorgement of the brain and the rapid softening that follows the effect of pathological congestion or of circulation interrupted by death? Lastly, we should perhaps take into account what Hunter called the 'stimulus of death', which triggers off the cessation of life without belonging to the disease on which it nevertheless depends [26]. In any case, the phenomena of exhaustion that occur at the end of chronic disease (muscular flaccidity, diminution of sensibility and conductibility) have more to do with a certain relationship between life and death than with a definite pathological structure.

Two series of questions confront a pathological anatomy that wishes to be based on a nosology: the first concerns the connexion between a temporal set of symptoms and a spatial coexistence of tissues; the second concerns death and the strict definition of its relation to life and disease. In its attempt to resolve these problems, Bichat's anatomy abandoned all its original meanings.

In order to overcome the first series of objections, there did not seem to be any need to modify the structure of the clinical gaze itself: was it not enough simply to observe the dead as one observes the living and to apply to corpses the diacritical principle of medi-cal observation: *the only pathological fact is a comparative fact*?

In their application of this principle Bichat and his successors found themselves in the company not only of Cabanis and Pinel, but also of Morgagni, Bonet, and Valsalva. The first anatomists knew very well that one had to be 'practised in the dissection of healthy bodies' if one wished to detect a disease in a corpse: otherwise, how could one distinguish an intestinal disease from those 'polypous concretions' that are caused by death or that sometimes affect the healthy at certain seasons? [27] One must also compare subjects who have died of the same disease, thus accepting the old principle already formulated by the *Sepulchretum* that alterations observed on all bodies define, if not the cause, at least the seat of the disease and perhaps its nature; those that differ from one autopsy to another are the result of effect, sympathy, or complication [28]. And finally, one must consider the comparison between what one sees of an altered organ and what one knows of its normal functioning: one must 'constantly compare these sensible phenomena that are proper to the health of each organ with the disorders of each of them present in its lesion' [29].

But the peculiarity of anatomo-clinical experience lies in having applied the diacritical principle to a much more complex and problematic dimension: that in which the recognizable forms of pathological history and the visible elements that it reveals on completion are articulated. Corvisart dreamt of replacing the old treatise of 1760 with the first definitive book of pathological anatomy, entitled *De sedibus et causis morborum per signa diagnostica investigatis et per anatomen confirmatis* [30]. And this anatomo-clinical coherence, which Corvisart perceived as a confirmation of nosology by autopsy, was defined by Laënnec in an opposite direction, as a rise of the lesion to the symptoms that it caused:

> Pathological anatomy is a science whose aim is the knowledge of the visible alterations produced on the organs of the human body by the state of disease. The opening up of corpses is the means of acquiring this knowledge; but in order for it to become of direct use . . . it must be joined to observation of the symptoms or alterations of functions that coincide with each kind of alteration in the organs [31].

The medical gaze must therefore travel along a path that had not so far been opened to it: vertically from the symptomatic surface to the tissual surface; in depth, plunging from the manifest to the hidden; and in both directions, as it must continuously travel if one wishes to define, from one end to the other, the network of essential necessities. The medical gaze, which, as we have seen, was directed upon the two-dimensional areas of tissues and symptoms, must, in order to reconcile them, itself move along a third dimension. In this way, anatomo-clinical range will be defined.

The gaze plunges into the space that it has given itself the task of traversing. In its primary form, the clinical reading implied an external, deciphering subject, which, on the basis of and beyond that which it spelt out, ordered and defined kinships [32]. In anatomo-clinical experience, the medical eye must see the illness spread before it, horizontally and vertically in graded depth, as it penetrates into the body, as it advances into its bulk, as it circumvents or lifts its masses, as it descends into its depths. Disease is no longer a bundle of characters disseminated here and there over the surface of the body and linked together by statistically observable concomitances and successions; it is a set of forms and deformations, figures, and accidents and of displaced, destroyed, or modified elements bound together in sequence according to a geography that can be followed step by step. It is no longer a pathological species inserting itself into the body wherever possible; it is the body itself that has become ill.

At first sight, it might be thought that this constitutes a reduction of the distance between the knowing subject and the object of knowledge. Did not the seventeenth- and eighteenth-century doctor remain 'at a distance' from his patient? Did he not observe him from afar, noting only the superficial, immediately visible marks and watching for phenomena, without physical contact or auscultation, guessing at the inside by external notations alone? Was not the change in medical knowledge at the end of the eighteenth century based essentially on the fact that the doctor came close to the patient, held his hand, and applied his ear to the patient's body, that by thus changing the balance, he began to perceive what was immediately behind the visible surface, and that he was thereby led gradually 'to pass on to the other side', and to map the disease in the secret depths of the body?

This amounts to no more than a minimal interpretation of the change. But one must not be misled by its theoretical discretion. It also involved a number of requisites, or references, that still have received very little attention: progress in observation, a wish to develop and extend experiment, an increasing fidelity to what can be revealed by sense-perceptible data, abandonment of theories and systems in favour of a more genuinely scientific empiricism. And behind all this, one supposes that the subject and object of knowledge remained what they were: their greater proximity and better adjustment simply made it possible for the object to reveal its own secrets with greater clarity or detail and for the subject to dispense with illusions that were an obstacle to truth. Established once and for all and placed definitively opposite one another, they could not but come closer to one another, reduce their distance, remove the obstacles that separated them, and discover the form of a reciprocal adjustment in the course of a historical transformation.

But this is surely a project on history, an old theory of knowledge whose effects and misdeeds have long been known. A more precise historical analysis reveals a quite different principle of adjustment beyond these adjustments: it bears jointly on the type of objects to be known, on the grid that makes it appear, isolates it, and carves up the elements relevant to a possible epistemic knowledge (*savoir*), on the position that the subject must occupy in order to map them, on the instrumental mediations that enables it to grasp them, on the modalities of registration and memory that it must put into operation, and on the forms of conceptualization that it must practice and that qualify it as a subject of legitimate knowledge. What is modified in giving place to anatomo-clinical medicine is not, therefore, the mere surface of contact between the knowing subject and the known object; it is the more general arrangement of knowledge that determines the reciprocal positions and the connexion between the one who must know and that which is to be known. The access of the medical gaze into the sick body was not the continuation of a movement of approach that had been developing in a more or less regular fashion since the day when the first doctor cast his somewhat unskilled gaze from afar on the body of the first patient; it was the result of a recasting at the level of epistemic knowledge (*savoir*) itself, and not at the level of accumulated, refined, deepened, adjusted knowledge (*connaissances*).

Whether it is as a result of an event that affected the arrangement of epistemic knowledge (*savoir*), proof of it is to be found in the fact that knowledge (*connaissances*) in the order of anatomo-clinical medicine is not formed in the same way and according to the same rules as in the mere clinic. It is not a matter of the same game, somewhat improved, but of a quite different game. Here are some of these new rules.

For the method of systematic identities, the anatomo-clinical substitutes what might be called a *chequered* or stratified analysis. The manifest repetitions often leave in a confused state morbid forms whose diversity can only be demonstrated by anatomy. The feeling of suffocating, sudden palpitations, especially after physical effort, quick, difficult breathing, waking up with a start, cachectic pallor, a feeling of pressure and constriction in the precordial region and of heaviness and numbness in the left arm are overwhelming signs of heart diseases in which only anatomy can distinguish pericarditis (which affects the investing membrane), aneurism (affecting the muscular tissue), or contractions and hardening (in which the heart is affected in its tendinous or fibrous parts) [33]. The coincidence, or at least the regular succession, of catarrh and phthisis does not prove that they are identical, despite the nosographers, since autopsy shows in one case an infection of the mucous membrane and in the other an alteration of the parenchyma, possibly to the point of ulceration [34]. But, inversely, two diseases like tuberculosis and haemoptysis, in which a symptomatology like that of Sauvages failed to find a sufficient link of frequency to group them together, must be placed together as belonging to the same local cell. The coincidence that defines pathological identity will be of value only for a locally isolated perception.

In other words, medical experience will substitute the *localization of the fixed point* for the *recording of frequencies*. The symptoms of pulmonary phthisis include coughing, difficulty in breathing, marasmus, hectic fever, and sometimes purulent expectoration; but none of these visible modifications is absolutely indispensable (there are tubercular patients who do not cough), and the order of their appearance is not strict (fever may appear early on or only towards the end of the evolution of the disease). There is only one constant phenomenon, the necessary and sufficient condition for the presence of phthisis: lesion of the pulmonary parenchyma, which, at autopsy, 'is shown to be dotted to a greater or lesser extent with purulent areas. In certain cases, they are so numerous that the lung seems to be no more than an alveolar tissue containing them. These areas are traversed by a large number of ridges; in the neighbouring parts one finds a certain degree of hardening' [35]. Above this fixed point, the symptoms slip and disappear; the index of probability that the clinic provided them with tends to be replaced by a single necessary implication that relates not to temporal frequency but to local constancy: 'Individuals must be regarded as phthisic who are neither feverous, nor

thin, nor suffering from purulent expectoration; it is enough that
the lungs should be affected by a lesion that tends to disorganize
and ulcerate them; phthisis is simply that lesion' [36].

Attached to that fixed point, the *chronological series* of symptoms is ordered, in the form of secondary phenomena, according
to the *ramification of the lesional space* and the logic that is peculiar
to it. Studying the 'strange and inexplicable' progress of certain
fevers, Petit makes a systematic comparison of the observations of
the disease and the result of autopsies: the succession of intestinal,
gastric, feverish, glandular, and even encephalic signs must be
originally attached as a whole to 'perfectly similar alterations of the
intestines'. The ileo-caecal valve is always covered with dark-red
stains and is swollen on the inside, and the glands of the corresponding mesenteric segment are swollen, dark-red and bluish in colour,
and deeply inflamed and congested. If the disease has lasted a long
time, there is ulceration and destruction of the intestinal tissue. It
can be admitted, therefore, that a deleterious action has taken place
in the digestive tract, whose functions are the first to be affected;
this agent is 'transmitted by absorption to the glands of the mesentery and to the lymphatic system' (hence the vegetative disorder),
and from there 'to the system as a whole', especially to its encephalic and nervous elements, which explains somnolence, the dead-
ening of the sense functions, delirium, and the phases of the
comatose state [37]. The succession of forms and symptoms then
appears simply as the chronological image of a more complex network: a spatio-temporal proliferation spreading from an original
attack throughout the entire organism.

The analysis of the anatomo-clinical perception reveals, therefore, three references (those of localization, site, and origin) that
modify the essentially temporal reading of the clinic. The organic
'cross-ruling' that makes it possible to determine fixed but arbores-
cent points does not abolish the density of pathological history to
the advantage of the pure anatomical surface; it introduces it into
the specified volume of the body, bringing about for the first time
in medical thought a coincidence of the morbid time and the
mappable course of organic masses. Then, but only then, pathological anatomy re-discovers the themes of Morgagni and, beyond him,
of Bonet: an autonomous organic space, with its own dimensions,
ways, and articulations, duplicates the natural or significative space
of nosology, and requires, essentially, that it should be brought
back. Born of the clinical concern to define the *structures of patho-*

logical kinship (cf. the *Traité des membranes*), the new medical perception finally attributed to itself the task of mapping the *figures of localization* (cf. the researches of Corvisart and G.-L. Bayle). The notion of *seat* has finally replaced that of *class*: 'What is observation', Bichat was already asking, 'if one is ignorant of the seat of the disease?' [38] And Bouillaud was to reply: 'If there is an axiom in medicine it is certainly the proposition that there is no disease without a seat. If one accepted the contrary opinion, one would also have to admit that there existed functions without organs, which is a palpable absurdity. The determination of the seat of disease or their localization is one of the finest conquests of modern medicine' [39]. Tissual analysis, whose original meaning was generic, could not fail, by its own structure, to assume very rapidly the value of a rule of localization.

Yet Morgagni was not re-discovered without a major modification. He had linked the notion of pathological seat with that of cause—*De sedibus et causis*; in the new pathological anatomy the determination of the seat did not involve an assignation of causality: the fact of finding ileo-caecal lesions in adynamic fevers is not a statement of determinant cause; Petit was to think of a 'deleterious agent' and Broussais of an irritation. This hardly mattered: to localize was to fix only a spatial and temporal starting point. For Morgagni, the seat was the point of insertion in the organism of the chain of causalities; it was identified with its ultimate link. For Bichat and his successors, the notion of seat is freed from the causal problematic (and in this respect, they are the heirs of the clinicians); it is directed towards the future of the disease rather than to its past; the seat is the point from which the pathological organization radiates. Not the *final cause*, but the *original site*. It is in this sense that the fixation onto a corpse of a segment of immobile space may resolve the problems presented by the temporal developments of a disease.

In eighteenth-century medical thought death was both the absolute fact and the most relative of phenomena. It was the end of life and, if it was in its nature to be fatal, it was also the end of the disease; with death, the limit had been reached and truth fulfilled, and by the same breach: in death, disease reached the end of its course, fell silent, and became a thing of memory. But if the traces of the disease happened to bite into the corpse, then no evidence could distinguish absolutely between what belonged to it and what to death; their signs intersected in indecipherable disorder. Death was that absolute beyond which there was neither life nor disease, but its disorganizations were like all morbid phenomena. In its original form, clinical experience did not call into question this ambiguous concept of death.

Pathological anatomy, the technique of the corpse, had to give this notion a more rigorous, that is, a more instrumental status. This conceptual mastery of death was first acquired, at a very elementary level, by the organization of clinics. The possibility of opening up corpses immediately, thus reducing to a minimum the latency period between death and the autopsy, made it possible for the last stage of pathological time and the first stage of cadaveric time almost to coincide. The effects of organic decomposition were virtually suppressed, at least in their most manifest, most disturbing form, so that the moment of death may act as a marker without density that rediscovers nosographical time, as the scalpel does organic space. Death is now no more than the vertical, absolutely thin line that joins, in dividing them, the series of symptoms and the series of lesions.

On the other hand, Bichat, taking up various suggestions made by Hunter, tried to distinguish between two types of phenomena that Morgagni's anatomy had confused: manifestations contemporary with the disease and those prior to death. In fact, an alteration need not refer to the disease and the pathological structure; it may refer to a different process, partly autonomous, partly dependent, that announces the coming of death. Thus muscular flaccidity belongs to the semiology of certain paralyses that are encephalic in origin, or of a vital affection such as asthenic fever; but one may also meet it in any chronic disease, or even in any acute episode, providing they are of sufficiently long duration; examples can be seen in inflammations of the arachnoid, or in the last stages of phthisis. The phenomenon, which would not have taken place without the disease, is not, however, the disease itself: it duplicates its duration with an evolution that indicates not a figure of the pathological, but the proximity of death; it designates, beneath the morbid process, the associated, but different process of 'mortification'.

These phenomena may well be similar in content to the fatal or favourable 'signs', so often analysed since Hippocrates. In structure and semantic value, however, they are very different: the sign referred to a possible outcome, by anticipation in time; and it indicated either the essential gravity of the disease, or its accidental gravity (whether due to a complication or to a therapeutic error). The phenomena of partial or progressive death prejudge no future: they show a process fulfilling itself; after apoplexy, most of the animal functions are naturally suspended, and consequently death has already begun for them, whereas the organic functions continue their own life [40]. Furthermore, the stages of this moving death do not follow only, or very much, the nosological forms, but, rather, the lines of facilitation proper to the organism. These processes indicate only in an incidental way the fatality of the disease; they speak of the permeability of life by death: when a pathological state is prolonged, the first tissues to be affected by mortification are always those in which nutrition is most active (the mucous membranes), followed by the parenchyma of the organs, and, in the final stage, by the tendons and aponeuroses [41].

Death is therefore multiple, and dispersed in time: it is not that absolute, privileged point at which time stops and moves back; like disease itself, it has a teeming presence that analysis may divide into time and space; gradually, here and there, each of the knots breaks, until organic life ceases, at least in its major forms, since long after the death of the individual, minuscule, partial deaths continue to dissociate the islets of life that still subsist [42]. In natural death, the animal life is extinguished first: first sensorial extinction, then the slowing down of brain activity, the weakening of locomotion, rigidity of the muscles and diminution of their contractility, quasi-paralysis of the intestines, and finally immobilization of the heart [43]. To this chronological picture of successive death must be added the spatial picture of the interactions that trigger off chain deaths throughout the organism. These occur in three main relays: heart, lungs, and brain. It may be established that the death of the heart does not involve the death of the brain through the nervous system but through the arterial network (cessation of the movement that sustains cerebral life) or through the vascular network (cessation of the movement, or on the contrary, the reflux, of black blood that obstructs the brain, compresses it, and prevents it from acting). It can also be shown how the death of the lung involves that of the heart, either because the blood has met a mechanical obstacle to circulation or because, by ceasing to act, the chemical reactions are deprived of food and the contraction of the heart is interrupted [44].

The processes of death, which can be identified neither with those of life nor with those of disease, are nevertheless of a nature to illuminate organic phenomena and their disturbances. The slow, natural death of the old man resumes in inverse direction the development of life in the child, in the embryo, perhaps even in the plant: 'The state of the animal that natural death annihilates is close to that in which it found itself inside its mother, and even to that of the vegetable that lives only within itself and for whom all nature is silent' [45]. The successive envelopes of life are detached naturally, enunciating their autonomy and truth in the very thing they deny. The system of functional dependencies and normal or pathological interactions is also illuminated by the detailed analysis of these deaths. It can be recognized that although there is direct action of the lung upon the heart, the heart is only indirectly influenced by the brain: apoplexy, epilepsy, narcotism, cerebral disturbances provoke no immediate, corresponding modification in the heart; only secondary effects may be produced through the mediation of muscular paralysis, interruption in breathing, or circulatory disorders [46]. Thus fixed in its own mechanisms, death, with its organic network, can no longer be confused with the disease or with its traces; on the contrary, it acts as a point of view on the pathological, and makes it possible to fix its forms and stages. In studying the causes of phthisis, G.-L. Bayle no longer considered death as a screen (functional or temporal) separating it from the disease, but as a spontaneous experimental situation providing access to the very truth of the disease, and to its different chronological phases. In fact, death may occur at any time in the pathological calendar, as a result either of the disease itself, of some additional affection, or of an accident. Once the non-variable phenomena and the variable manifestations of death are known and mastered, one may reconstitute, by means of this opening onto time, the evolution of a whole morbid series. For phthisis, these are, first of all, firm, homogeneous, whitish tubercles; then softer formations, including at the centre a nucleus of purulent matter that changes colour; finally, a state of suppuration causing ulcers and a destruction of the pulmonary parenchyma [47]. Systematizing the same method, Laënnec was able to show, against Bayle himself, that melanosis did not constitute a distinct pathological type but a possible phase of evolution. The time of death may slide along the entire length of the morbid evolution; and as this death loses its opaque character, it becomes, paradoxically, and by virtue of its effect of temporal interruption, the instrument by which the duration of the disease can be integrated with the immobile space of a dissected body.

Life, disease, and death now form a technical and conceptual trinity. The continuity of the age-old beliefs that placed the threat of disease in life and of the approaching presence of death in disease is broken; in its place is articulated a triangular figure the summit of which is defined by death. It is from the height of death that one can see and analyse organic dependences and pathological sequences. Instead of being what it had so long been, the night in which life disappeared, in which even the disease becomes blurred, it is now endowed with that great power of elucidation that dominates and reveals both the space of the organism and the time of the disease. The privilege of its intemporality, which is no doubt as old as the consciousness of its imminence, is turned for the first time into a technical instrument that provides a grasp on the truth of life and the nature of its illness. Death is the great analyst that shows the connexions by unfolding them, and bursts open the wonders of genesis in the rigour of decomposition: and the word *decomposition* must be allowed to stagger under the weight of its meaning. Analysis, the philosophy of elements and their laws, meets its death in what it had vainly sought in mathematics, chemistry, and even language: an unsupersedable model, prescribed by nature; it is on this great example that the medical gaze will now rest. It is no longer that of a living eye, but the gaze of an eye that has seen death—a great white eye that unties the knot of life.

There is much that might be said about Bichat's 'vitalism'. It is true that in trying to circumscribe the special character of the living phenomenon Bichat linked to its specificity the risk of disease: a simply physical body cannot deviate from its natural type [48]. But this does not alter the fact that the analysis of the disease can be carried out only from the point of view of death—of the death which life, by definition, resists. Bichat relativized the concept of death, bringing it down from that absolute in which it appeared as an indivisible, decisive, irrecoverable event: he volatilized it, distributed it throughout life in the form of separate, partial, progressive deaths, deaths that are so slow in occurring that they extend even beyond death itself. But from this fact he formed an essential structure of medical thought and perception: that to which life is *opposed* and to which it is *exposed*; that in relation to which it is living *opposition*, and therefore *life*; that in relation to which it is analytically *exposed*, and therefore *true*. Magendie, and Buisson before him, went to the crux of the problem, but as biologists, when they criticized the definition of life with which the *Recherches physiologiques* opens: 'A false idea, since to die signifies in every language to cease to live, and so the supposed definition is reduced to the following vicious circle: Life is the totality of functions that resist the absence of life' [49]. But it was from one of his earliest experiences as an anatomo-pathologist that Bichat set out: an experience, which he himself had constituted, in which death was the only possibility of giving life a positive truth. The irreducibility of the living to the mechanical or chemical is secondary only in relation to the fundamental link between life and death. Vitalism appears against the background of this 'mortalism'.

A vast distance had been traversed since that relatively recent moment when Cabanis assigned to the knowledge of life the same origin and foundation as life itself:

> Nature intended that the source of our knowledge shall be the same as that of life. One must receive impressions in order to live; one must receive impressions in order to know; and since the need to study is always directly proportional to their action upon us, it follows that our means of instruction are always proportionate to our needs [50].

For Cabanis, as for the eighteenth century and for a whole tradition that was already familiar in the Renaissance, the knowledge of life was based on the essence of the living, since it, too, is no more than a manifestation of it. That is why one never attempted to conceive of disease on the basis of the living, or of its (mechanical) models and (humoral, chemical) constituents; vitalism and anti-vitalism both sprang from this fundamental anteriority of life in the experience of disease. With Bichat, knowledge of life finds it origin in the destruction of life and in its extreme opposite; it is at death that disease and life speak their truth: a specific, irreducible truth, protected from all assimilations to the inorganic by the circle of death that designates them for what they are. Cabanis, who thrust life back so far into the depths of origins, was naturally more mechanistic than Bichat, who conceived of it only in relation to death. From the Renaissance to the end of the eighteenth century, the knowledge of life was caught up in the circle of life folded back upon and observing itself; from Bichat onwards it is 'staggered' in relation to life, and separated from it by the uncrossable boundary of death, in the mirror of which it observes itself.

It was no doubt a very difficult and paradoxical task for the medical gaze to operate such a conversion. An immemorial slope as old as men's fear turned the eyes of doctors towards the elimination of disease, towards cure, towards life: it could only be a matter of restoring it. Behind the doctor's back, death remained the great dark threat in which his knowledge and skill were abolished; it was the risk not only of life and disease but of knowledge that questioned them. With Bichat, the medical gaze pivots on itself and demands of death an account of life and disease, of its definitive immobility of their time and movements. Was it not necessary that medicine should circumvent its oldest care in order to read, in what provided evidence of its failure, that which must found its truth?

But Bichat did more than free medicine of the fear of death. He integrated that death into a technical and conceptual totality in which it assumed its specific characteristics and its fundamental value as experience. So much so that the great break in the history of Western medicine dates precisely from the moment clinical experience became the anatomo-clinical gaze. Pinel's *Médecine clinique* dates from 1802; *Les Révolutions de la Médecine* appeared in 1804; the rules of analysis seem to triumph in the pure decipherment of symptomatic totalities. But a year before, Bichat had relegated them to history:

> for twenty years, from morning to night, you have taken notes at patients' bedsides on affections of the heart, the lungs, and the gastric viscera, and all is confusion for you in the symptoms which, refusing to yield up their meaning, offer you a succession of incoherent phenomena. Open up a few corpses: you will dissipate at once the darkness that observation alone could not dissipate [51].

The living night is dissipated in the brightness of death.

NOTES

[1] P. Rayer, *Sommaire d'une histoire abrégée de l'anatomie pathologique* (Paris, 1818, introduction, p. v).

[2] Rostan, *Traité élémentaire de diagnostic, de pronostic, d'indications thérapeutiques* (Paris, 1826, vol. I, p. 8).

[3] J.-L. Alibert, *Nosologie naturelle* (Paris, 1817, Préliminaire, I, p. lvi).
[4] Cf. the account of the autopsy of the giant in D. Ottley, 'Vie de John Hunter', in J. Hunter, *Oeuvres complètes* (Fr. trans., Paris, 1839, vol. I, p. 126).
[5] M.-A. Petit, 'Éloge de Desault' (1795), in *Médecine due Coeur*, p. 108.
[6] Cf. Gilibert, *L'Anarchie médicinale* (Neuchâtel, 1772, vol. I, p. 100).
[7] X. Bichat, *Traité des membranes* (edn. of 1827, with notes by Magendie), p. 6.
[8] *Ibid.*, p. 1.
[9] *Ibid.*, pp. 6–8.
[10] *Ibid.*, p. 2.
[11] *Ibid.*, p. 5.
[12] *Anatomie pathologique* (Paris, 1825, p. 3).
[13] *Anatomie générale* (Paris, 1801, vol. I, avant-propos, p. xcvii).
[14] *Anatomie pathologique*, p. 39.
[15] *Traité des membranes*, pp. 213–64.
[16] *Anatomie pathologique*, p. 12.
[17] Quoted in Lallemand, *Recherches anatomo-pathologiques sur l'encéphale* (Paris, 1820, vol. I, p. 88).
[18] *Anatomie générale* (Paris, 1801, vol. I, p. lxxix).
[19] Pinel, *Nosographie philosophique*, I, p. xxviii.
[20] *Anatomie générale*, vol. I, pp. xcvii–xcviii.
[21] R. Laënnec, *Dictionnaire des Sciences médicales*, article 'Anatomie pathologique', II, p. 49.
[22] *Ibid.*, pp. 450–2.
[23] Alibert, *op. cit.*, avertissement, p. ii; cf. other classifications based on pathological anatomy in Marandel, *Essai sur les irritations* (Paris, 1807), or in Andral.
[24] F.-J. Double, *Séméiologie générale*, vol. I, pp. 56–7.
[25] *Ibid.*, pp. 64–7.
[26] J. Hunter, *Oeuvres complètes* (Paris, 1839, vol. I, p. 262).
[27] Morgagni, 'Recherches anatomiques', *Encyclopédie des Sciences médicales*, 7th section, vol. VII, p. 17.
[28] Th. Bonet, *Sepulchretum*, preface; this principle is recalled by Morgagni, *op. cit.*, p. 18.
[29] Corvisart, *Essai sur les maladies et les lésions organiques, du coeur et des gros vaisseaux* (Paris, 3rd edn., 1818, discours préliminaire, p. xii).
[30] *Ibid.*, p. v.
[31] Laënnec, *op. cit.*, p. 47.
[32] Cf. Chapter 7.
[33] Corvisart, *op. cit.*
[34] G.-L. Bayle, *Recherches sur la phthisie pulmonaire* (Paris, 1810).
[35] X. Bichat, *Anatomie pathologique* (Paris, 1825, p. 174).
[36] Bayle, *op. cit.*, pp. 8–9.
[37] M.-A. Petit, *Traité de la fièvre entéro-mésentérique* (Paris, 1813, especially pp. xix, xxx, and 132–41).
[38] Bichat, *Anatomie générale*, vol. I, p. xcix.
[39] Bouillaud, *Philosophie médicale*, p. 259.

[40] X. Bichat, *Recherches physiologiques sur la vie et la mort* (Magendie, edn. p. 251).

[41] Bichat, *Anatomie pathologique*, p. 7.

[42] Bichat, *Recherches physiologiques*, p. 242.

[43] *Ibid.*, pp. 234 and 238.

[44] *Ibid.*, pp. 253 and 538.

[45] *Ibid.*, p. 238.

[46] *Ibid.*, pp. 480 and 500.

[47] Bayle, *op. cit.*, pp. 21–4.

[48] Cf. G. Canguilhem, *La connaissance de la vie* (Paris, 1952, p. 195).

[49] F. R. Buisson, *De la division la plus naturelle des phénomènes physiologiques* (Paris, 1802, p. 57). Cf. also Magendie, n. 1, p. 2 of his edition of *Recherches physiologiques*.

[50] Cabanis, *Du degré de certitude de la médecine* (3rd edn., Paris, 1819, pp. 76–7).

[51] Bichat, *Anatomie générale*, avant-propos, p. xcix.

9 · The Visible Invisible

From the point of view of death, disease has a land, a mappable territory, a subterranean, but secure place where its kinships and its consequences are formed; local values define its forms. Paradoxically, the presence of the corpse enables us to perceive it living —living with a life that is no longer that of either old sympathies or the combinative laws of complications, but one that has its own roles and its own laws.

I. PRINCIPLE OF TISSUAL COMMUNICATION

Roederer and Wagler had already defined *morbus mucosus* as an inflammation that may affect both the internal and the external surface of the alimentary canal throughout its full length [1]. Bichat generalized this observation: a pathological phenomenon follows in the organism the privileged way prescribed by tissual identity. Each type of membrane has its own pathological modalities: 'Since diseases are merely alterations of vital properties, and since each tissue differs from others in relation to these properties, it is evident that it must also differ in its diseases' [2]. The arachnoid may be affected by the same forms of dropsy as the pleura of the lung or the peritoneum, since there are serous membranes present in each case. The network of sympathies that was fixed only on unsystematized resemblances, empirical observations, or a conjectural assignation of the nervous network now rests on a strict analogy of structure: when the envelopes of the brain are inflamed, the sensitivity of the eyes and ears is sharpened; in the operation of hydrocele by injection, the irritation of the vaginal wall causes pains in the lumbar region; an inflammation of the intestinal pleura may, by a 'sympathy of tonicity', cause a cerebral affection [3]. The pathological course now has its obligatory ways.

II. PRINCIPLE OF TISSUAL IMPERMEABILITY

This is the correlative of the preceding principle. Extending in areas, the morbid process follows a tissue horizontally, without penetrating vertically into others. Sympathetic vomiting concerns the fibrous tissue, not the mucous membrane of the stomach; diseases of the periosteum are alien to bone, and when there is catarrh in the bronchi, the pleura remains intact. The functional unity of an organ is not enough to force the communication of a pathological fact from one tissue to another. In hydrocele, the testicle remains intact in the midst of inflammation of the enveloping tunic [4]; while infections of the cerebral pulp are rare, those of the arachnoid are frequent, and of a very different type, again, from those of the pia mater. Each tissual stratum possesses and retains its own pathological characteristics. Morbid diffusion is a matter of isomorphic surfaces, not of proximity or of superposition.

III. PRINCIPLE OF PENETRATION BY BORING

Without calling them into question, this principle limits the preceding two. It compensates the rule of homology by the rules of regional influences, and the rule of impermeability by admitting forms of penetration by layer. An affection may last sufficiently long to impregnate subjacent or neighbouring tissues: this is what occurs in chronic diseases like cancer, when all the tissues of an organ are successively affected and, in the end, are 'confused in a common mass' [5]. Less easily assignable movements also occur: not by impregnation or by contact but by a double movement from one tissue to another, and from a structure to a function. The alteration of one membrane may, without affecting the neighbouring membrane, prevent more or less completely the performance of its functions: the mucous secretions of the stomach may be affected by inflammation of the fibrous tissues; and the intellectual functions may be affected by lesions of the arachnoid [6]. The forms of intertissual penetration may be even more complex: in affecting the investing membrane of the heart, pericarditis may cause a functional disorder resulting in hypertrophy of the organ, and therefore a modification of its muscular substance [7]. At its origin, pleurisy concerns only the pleura of the lung; but as a result of the disease, the pleura may secrete an albuminous liquid which, in chronic cases, covers the whole lung; the lung atrophies, and its activity is diminished to the point of an almost total cessation of its functioning, and it is then so reduced in surface and volume that it seems as if most of its tissue has been destroyed [8].

IV. PRINCIPLE OF THE SPECIFICITY OF THE MODE OF ATTACK ON THE TISSUES

Alterations whose trajectory and work are determined by the preceding principles belong to a typology that depends not only on the point that they attack but on their own nature. Bichat did not go very far in the description of these various modes, since he distinguished only between inflammations and scirrhi. Laënnec, as we have seen [9], attempted a general typology of alterations (of texture, of form, of nutrition, of position, and those due to the presence of foreign bodies). But the very notion of an alteration of texture is inadequate to describe the various ways in which a tissue may be attacked in its internal constitution. Dupuytren proposed to distinguish between transformations from one tissue to another and the productions of new tissues. In one case, the organism produces a tissue that exists regularly but that is usually found only in another localization, in the case of unnatural ossifications; cellular, adipose, fibrous, cartilaginous, osseous, serous, synovial, and mucous productions may be enumerated; such cases are *aberrations* of the laws of life, not *alterations*. In the contrary case, in which a new tissue is created, the laws of organization have been fundamentally disturbed; the lesional tissue is different from any tissue existing in nature; inflammation, tubercles, scirrhi, and cancer are of this kind. Finally, articulating this typology onto the principles of tissual localization, Dupuytren noted that each membrane has its special type of alteration: for example, polyps on the mucous membranes or dropsy in the serous membranes [10]. It was by applying this principle that Bayle was able to follow the evolution of phthisis from beginning to end, recognize the unity of its processes, specify its

forms, and distinguish it from affections whose symptomatology
may be similar but which belong to an absolutely different type of
alteration. Phthisis is characterized by a 'progressive disorganization'
of the lung, which may assume a tuberculous, ulcerous, calculous,
granulous, melanotic, or cancerous form; and it must be confused
neither with irritation of the mucous membranes (catarrh), nor
with alteration of the serous secretions (pleurisy), nor, above all,
with an alteration that also attacks the lung itself, but in the form
of inflammation, namely, chronic pleuropneumonia [11].

V. PRINCIPLE OF ALTERATION OF ALTERATION

Generally speaking, the preceding rule excludes the diagonal affections that intersect various modes of attack and use them in
turn. However, there are effects of facilitation that link different
disorders together: inflammation of the lungs and catarrh do not
constitute tuberculosis, but they do encourage its development [12].
Chronicity, or at least the persistence of an attack over a period of
time, sometimes permits one affection to take over from another. In
a sudden type of fluxion, cerebral congestion causes a distension of
the vessels (hence vertigo, dizziness, optical illusions, ringing in
the ears) or, if it is concentrated in one point, a rupture of the vessels
with resulting haemorrhage or immediate paralysis. But if the
congestion occurs by means of a slow invasion, there is first a san-
guineous infiltration into the cerebral matter (accompanied by convulsions and pains), a corresponding softening of this substance—
which, by admixture with the blood, alters in depth and agglutinates
to form inert islets (hence paralyses)—and finally a complete disorganization of the arteriovenous system in the cerebral parenchyma
and often even in the arachnoid. From the appearance of the earliest
forms of softening, serous discharges and then an infiltration of pus
that sometimes gathers into an abscess can be observed: finally, the
suppuration and extreme softening of the vessels replace the irritation due to their congestion and hypertension [13].

These principles define the rules of the pathological cursus and
describe in advance the possible paths that it must follow. They
fix the network of its space and development, revealing in transparency the nervures of the disease. The disease assumes the figure
of a great organic vegetation, which has its own forms of sprouting,
its own ways of taking root, and its own privileged regions of

growth. Spatialized in the organism in accordance with their own lines and areas, pathological phenomena take on the appearance of living processes. This has two consequences: disease is hooked onto life itself, feeding on it, and sharing in that 'reciprocal commerce of action in which everything follows everything else, everything is connected with everything else, everything is bound together' [14]. It is no longer an event or a nature imported from the outside; it is life undergoing modification in an inflected functioning: 'In the final analysis, every pathological phenomenon derives from their augmentation, diminution, and alteration' [15]. Disease is a deviation within life. Furthermore, each morbid group is organized according to the model of a living individual: there is a life of tubercles and a life of cancers. There is a life of inflammation; the old rectangle that qualifies it (tumour, redness, heat, pain) is inadequate to restore its development throughout the various organic stratifications: in the blood capillaries, it is conveyed by resolution, gangrene, induration, suppuration, and abscess; in the white capillaries, the curve moves from resolution to white, tuberculous suppuration, and from there to incurable rodent ulcers [16]. So the idea of a disease attacking life must be replaced by the much denser notion of *pathological life*. Morbid phenomena are to be understood on the basis of the same text of life, and not as a nosological essence: 'Diseases have been regarded as a disorder; one has failed to see in them a series of phenomena all dependent upon one another, usually tending to a particular end: pathological life has been completely neglected.'

Is this, at last, a non-chaotic, ordered development of disease? But it had already been a long-acquired fact; botanical regularity, the constancy of clinical forms had brought order to the world of illness long before the advent of the new anatomy. It was not the fact of ordering that was new, but its mode and basis. Between Sydenham and Pinel disease assumed a source and a face in a general structure of rationality concerning *nature* and the order of things. From Bichat onwards, the pathological phenomenon was perceived against the background of *life*, thus finding itself linked to the concrete, obligatory forms that it assumed in an organic individuality. Life, with its finite, defined margins of variation, was to play the same role in pathological anatomy as the broad notion of nature played in nosology: it was the inexhaustible, but closed basis in which disease finds the ordered resources of its disorders. A distant, theoretical change that, in the long term, modified a philosophical horizon; but can it be said that it affected at once a world of perception and the gaze that a doctor turns upon a patient?

It did so, no doubt, in a very considerable, decisive way. The phenomena of disease find there their ontological support. Paradoxically, clinical 'nominalism' left floating at the limit of the medical gaze, at the grey frontiers of the visible and invisible, something that was both the totality of phenomena and their law, their point of recollection, as well as the strict rule of their coherence; disease had truth only in symptoms, but it was symptoms given in truth. The discovery of the vital processes as the content of disease makes it possible to give a foundation that is nevertheless neither distant nor abstract: a foundation as close as possible to what is manifest; disease will now be merely the pathological form of life. The great nosological essences, which hovered over the order of life and threatened it, are now circumvented by it: life is the immediate, the present, and the perceptible *beyond* disease; and disease, in turn, finds its phenomena once more in the morbid form of life.

Is this the reactivation of a vitalist philosophy? It is true that the thought of Bordeu or Barthez was familiar to Bichat. But if vitalism is a schema of specific interpretation of healthy or morbid phenomena in the organism, it is much too feeble a concept to account for an event of the significance of the discovery of pathological anatomy. Bichat revived the theme of the specificity of the living only in order to place life at a deeper, more concealed ontological level: for him, it is not a set of characteristics that are distinguished from the inorganic, but the background against which the opposition between the organism and the non-living may be perceived, situated, and laden with all the positive values of conflict. Life is not the form of the organism, but the organism is the visible form of life in its resistance to that which does not live and which opposes it. An argument between vitalism and mechanism, or between humourism and solidism, had meaning only insofar as nature, too broad an ontological foundation, left room for the play of those interpretive models: normal or abnormal functioning could be explained only by reference either to a pre-existing form or to a specific type. But as soon as life explained not solely a series of natural figures but assumed sole responsibility for the role of the absolute, considered basis that the eighteenth century accorded to nature, the very idea of vitalism lost its signification and the essence of its content. By giving life, and pathological life, so fundamental a status, Bichat freed medicine from the vitalist and other related problems. Hence the feeling, which bore up the theoretical reflexion of most doctors at the beginning of the nineteenth century, that they were free at last of systems and speculations. The clinicians Cabanis and Pinel felt that their method was realized philosophy [17]; the anatomo-pathologists discovered in theirs a non-philosophy, an abolished philosophy, that they had conquered in learning at last to perceive: it was simply a question of a shift in the ontological foundation on which their perception was based. It seemed to them that an absolute theoretical reduction had taken place: a mirage effect due solely to a radical interpretation of life.

At this epistemological level, life is to be distinguished from the inorganic only at a superficial level, and in the order of its consequences. It is profoundly bound up with death, as to that which positively threatens to destroy its living force. In the eighteenth century, disease was both nature and counter-nature, since it possessed an ordered essence, but it was of its essence to compromise natural life. From Bichat onwards, disease was to play the same dual role, but between life and death. Let us be clear about this: an experience devoid of both age and memory knew, well before the advent of pathological anatomy, the way that led from health to disease, and from disease to death. But this relationship had never been scientifically conceived or structured in medical perception; at the beginning of the nineteenth century it acquired a figure that can be analysed at two levels. That which we know already: death as the absolute point of view over life and opening (in all senses of the term, even the most technical) on its truth. But death is also that against which life, in daily practice, comes up against; in it, the living being resolves itself naturally: and disease loses its old status as an accident, and takes on the internal, constant, mobile dimension of the relation between life and death. It is not because he falls ill that man dies; fundamentally, it is because he may die that man may fall ill. And beneath the chronological life/disease/death relation, another, earlier, deeper figure is traced: that which links life and death, and so frees, besides, the signs of disease.

Earlier, death appeared as the condition of the gaze that gathered together, in a reading of surfaces, the time of pathological events; it enabled the disease to be articulated at last in a true discourse. Now it appears as the source of disease in its very being, that possibility internal to life, but stronger than it, which exhausts it, diverts it, and finally makes it disappear. Death is disease made possible in life. And although it is true that for Bichat the pathological pheno-menon is connected with the physiological process and derives from it, this derivation, in the gap that it constitutes, and which denounces the morbid fact, is based upon death. Deviation in life is of the order of life, but of a life that moves towards death.

Hence the importance assumed with the appearance of patho-logical anatomy by the concept of 'degeneration'. It was already an old notion: Buffon applied it to individuals or series of individuals that diverged from their specific type [18]; doctors also used it to designate that weakening of natural robust humanity that life in society, civilization, laws, and language condemn little by little to a life of artificiality and disease; to degenerate was to describe a decline from an original status, figuring by natural right at the sum-mit of the hierarchy of perfections and times; in this notion is gathered up all that was most negative in the historical, the atypical, and the counter-natural. Based, from Bichat onwards, on a percep-tion of death that was at last conceptualized, degeneration was gradually to be given a positive content. At the frontier of the two significations, Corvisart defined organic disease by the fact that 'an organ, or any solid living thing, is as a whole or in one of its parts degenerated enough from its natural condition for its easy, regular, constant action to be endangered or disordered in a perceptible and permanent way' [19]. A broad definition that embraces every possible form of anatomical and functional alteration; and, again, a negative definition, since degeneration is merely a distance taken in relation to a state of nature: a definition that nevertheless author-izes the first movement of a positive analysis, since Corvisart speci-fies its forms as 'alterations of contexture', modifications of symmetry, and changes in 'the physical and chemical mode of being' [20]. In this sense, degeneration is the external curve in which lodge the singular points of pathological phenomena; at the same time it is the principle governing the reading of their fine structure.

Within such a general framework, the point of application of the concept was open to controversy. In a report on organic diseases, Martin [21] contrasted tissual formations (whether of a known or a new type) with degenerations, in the strict sense, which modify only the form or internal structure of the tissue. On the other hand, Cruveilhier, also criticizing too wide a use of the term 'degeneration', wished to reserve it for that disordered activity of the organism that creates tissues that have no parallel in the state of health; such tissues, which usually present 'a fatty, greyish texture', are to be found in tumours, in the irregular masses formed at the expense of the organs, in ulcers or fistulas [22]. According to Laënnec, one may speak of degeneration in two precise cases: when one tissue changes into another that exists in a different form and localization in the organism (osseous degeneration of the cartilages, fatty degeneration of the liver); and when a tissue assumes a texture or configuration that has no pre-existing model (tuberculous degeneration of the lymphatic glands or of the pulmonary parenchyma; scirrhous degeneration of the ovaries or testicles) [23]. But in any event one cannot speak of degeneration in the case of a pathological superposition of tissues. An apparent thickening of the dura mater is not always an ossification; in anatomical examination, it is possible to detach on the one hand the arachnoid and on the other the dura mater: a tissue is then revealed that has been deposited between the membranes, but this is not a degenerate development of one of them. One should speak of degeneration only in the case of a process that takes place within the tissual texture; it is the pathological dimension of its own evolution. A tissue degenerates when it is sick *qua* tissue.

This tissual sickness may be characterized by three indices. It is not simply a decline, nor is it a free deviation; it obeys certain laws: 'Nature is constrained by constant laws in the destruction as in the construction of beings' [24]. Organic legality is not, therefore, simply a precarious, delicate process; it is a reversible structure the stages of which follow a certain definite direction: 'the phenomena of life follow laws, even in their alterations' [25]. A direction indicated by figures whose level of organization becomes weaker and weaker; first, the morphology becomes blurred (irregular ossifications); then intra-organic differentiations occur (cirrhosis, hepatization of the lung); finally, the internal cohesion of the tissue disappears: when it is inflamed, the cellular sheath of the arteries 'allows itself to be cut like lard' [26], and the tissue of the liver may be pulled away with no effort. This disorganization may even become auto-destruction, as in the case of tuberculous degeneration, when the ulceration of the nuclei causes the destruction not only of the parenchyma but of the tubercles themselves. Degeneration is not, therefore, a return to the inorganic; or, rather, it is such a return only insofar as it is infallibly orientated towards death. The disorganization that characterizes it is not that of the non-organic, it is that of the non-living, of life caught up in the process of self-destruction: 'we must call pulmonary phthisis any lesion of the lung which, left to itself, produces a progressive disorganization of that organ as a result of which occur its alteration and, finally, death' [27]. That is why there is a form of degeneration that constantly accompanies life and, throughout its entire duration, defines its confrontation with death: 'The idea of the alteration and lesion of parts of our organs by the very fact of their action is one that most authors have not deigned to consider' [28]. Wear is an ineffaceable temporal dimension of organic activity: it measures the silent work that disorganizes tissues simply by virtue of the fact that they carry out their functions, and that they encounter 'a host of external agents' capable of 'overpowering their resistance'. Gradually, from the moment they move into action and confront the outside world, death begins to indicate its imminence: it insinuates itself not only in the form of possible accident; with life it forms its movements and times, the single web that both constitutes and destroys it.

Degeneration lies at the very principle of life, the necessity of death that is indissociably bound up with life, and the most general possibility of disease. A concept whose structural link with the anatomo-pathological method now appears in all its clarity. In anatomical perception, death was the point of view from the height of which disease opened up onto truth; the life/disease/death trinity was articulated in a triangle whose summit culminated in death; perception could grasp life and disease in a single unity only insofar as it invested death in its own gaze. And now the same configuration can be seen in perceived structures, but in an inverted mirror image: life with its real duration and disease as a possibility of deviation find their origin in the deeply buried point of death; it commands their existence from below. Death, which, in the anatomical gaze, spoke retroactively the truth of disease, makes possible its real form by anticipation.

For thousands of years, medicine had sought a mode of articulation that might define the relations between disease and life. Only the intervention of a third term was able to give to their encounter, to their coexistence, to their interferences, a form based both on conceptual possibility and on perceived plenitude; this third term is death. On the basis of death, disease is embodied in a space that coincides with that of the organism; it follows its lines and dissects it; it is organized in accordance with its general geometry; it is also inflected towards its singularities. From the moment death was introduced into a technical and conceptual organon, disease was able to be both spatialized and individualized. Space and individual, two associated structures deriving necessarily from a death-bearing perception.

In the depths of its being, disease follows the obscure, but necessary ways of tissual reactions. But what now becomes of its visible body, that set of phenomena without secrets that makes it entirely legible for the clinicians' gaze: that is, recognizable by its signs, but also decipherable in the symptoms whose totality defined its essence without residue? Does not the whole of this language incur the risk of being relieved of its specific weight and reduced to a series of surface events, lacking in both grammatical structure and semantic necessity? In assigning to disease silent paths in the enclosed world of bodies, pathological anatomy reduces the importance of clinical symptoms and substitutes for a methodology of the visible a more complex experience in which truth emerges from its inaccessible reserve only in the passage to the inert, to the violence of the dissected corpse, and hence to forms in which living signification withdraws in favour of a massive geometry.

A new reversal of the relations between signs and symptoms. In the earliest form of clinical medicine, the sign was not by nature different from symptoms [29]. Every manifestation of disease could, without essential modification, take on the value of a sign, providing an informed medical reading could place it in the chronological totality of the illness. Every symptom was a potential sign, and the sign was simply a read symptom. Now, in an anatomo-clinical perception the symptom may quite easily remain silent, and the significant nucleus with which one believed it to be armed prove to be non-existent. What visible symptom can indicate pulmonary phthisis with certainty? Neither difficulty in breathing, which may be found in a case of chronic catarrh, and not be found in a tubercular patient; nor coughing, which also belongs to neuro-pneumonia but not always to phthisis; nor hectic fever, which is frequent in pleurisy, but which often appears only in the latter stage of phthisis [30]. The silence of symptoms can be circumvented, but it cannot be overcome. The sign plays precisely this role of a detour: it is not an expressive symptom, but one which is substituted for the fundamental absence of expression in the symptom. In 1810, Bayle had been forced to reject in turn all the semeiological indications of phthisis: none was either evident or certain. Nine years later, Laënnec, sounding a patient whom he believed to be suffering from pulmonary catarrh, combined with bilious fever, had the impression that he was listening to the voice emerging directly out of the chest, and this on a small surface of about a square inch. Perhaps it was the effect of a pulmonary lesion, a sort of opening in the body of the lung. He met with the same phenomenon in about twenty consumptives; then he distinguished it from a fairly similar phenomenon to be observed in pleurisy patients: the voice also seemed to emerge from the chest, but it was more than naturally sharp; it seemed thin and quavering [31]. Laënnec therefore laid down 'pectoriloquy' as the only certain pathognomonic sign of pulmonary phthisis, and 'egophony' as the sign of pleuretic discharge. It can be seen that in anatomo-clinical experience the sign has an entirely different structure from that attributed to it, only a few years earlier, by the clinical method. In Zimmermann's or Pinel's perception, the sign was all the more eloquent, all the more certain, the more surface it occupied in the manifestations of the disease: thus fever was the major symptom, and consequently the most certain sign, and the one closest to the essential, by which the series of diseases bearing precisely the name of 'fever' could be recognized. For Laënnec, the value of the sign is no longer related to symptomatic extension; its marginal, restricted, almost imperceptible character enables it to traverse, diagonally as it were, the visible body of the disease (composed of general and uncertain elements) and to attain its nature at a stroke. By that very fact, it divests itself of the statistical structure that it possessed in pure clinical perception: in order for it to produce certainty, a sign had to belong to a convergent series, and it was the random configuration of the whole that bore the truth; now the sign speaks alone, and what it declares is apodictic: coughing, chronic fever, weakness, expectoration, and haemoptysis make phthisis more and more probable, but, in the last resort, never quite certain; pectoriloquy alone designates it without any possibility of error. Finally, the clinical sign referred to the disease itself, the anatomo-clinical sign to the lesion; and although certain tissue alterations are common to several diseases, the sign that reveals them can say nothing about the nature of the disorder: one may observe hepatization of the lung, but the sign that indicates it will not say what disease is responsible for that condition [32]. The sign, then, can refer only to a lesional occurrence, never to a pathological essence.

Significant perception is therefore structurally different in the world of the clinical as it existed in its first form, and as modified by the anatomical method. This difference is apparent even in the way in which the pulse was taken before and after Bichat. For Menuret, the pulse is a sign because it is a symptom, that is, insofar as it is a natural manifestation of the disease, and fully communicates with its essence. Thus a 'full, strong, rebounding' pulse indicates a plethora of blood, vigorous pulsations, and congestion of the vascular system, all of which suggest the possibility of a violent haemorrhage. The pulse 'holds by its causes to the constitution of the machine, to the most important and most extensive of its functions; by its skilfully grasped and developed characteristics, it uncovers the whole inside of man'; thanks to the pulse, 'the doctor shares in the science of the supreme being' [33]. In distinguishing between capital, pectoral, and ventral pulsations, Bordeu did not modify the form of perception of the pulse. It was still a question of reading a particular pathological state in the course of its evolution, and of foreseeing its most probable development; thus the simple pectoral pulse is soft, full, dilated; the pulsations are equal, but undulating, forming a sort of double wave 'with an ease, a softness, and a gentle force of oscillation that makes it impossible to confuse this kind of pulse with the others' [34]. It is the indication of an evacuation in the chest region. When Corvisart, on the other hand, takes his patient's pulse, it is not the symptom of an affection that he seeks, but the sign of a lesion. The pulse no longer possesses expressive value in its qualities of softness or fullness; but anatomo-clinical experience made it possible to draw up a picture of the bi-univocal correspondences between the appearance of the pulsations and each lesional type: the pulse is strong, hard, vibrant, and frequent in active aneurisms without complications; soft, slow, regular, easy to smother in simple passive aneurisms; irregular, unequal, undulating in permanent contractions; intermittent, irregular at intervals in temporary contractions; weak and scarcely perceptible in hardenings, ossifications, softenings; rapid, frequent, disordered, and almost convulsive in cases of the rupture of one or several bunches of fleshy fibres [35]. It is no longer a question of a science analogous with that of the Supreme Being, conforming to the laws of natural movements, but of the formulation of a certain number of perceptions of signals.

The sign no longer speaks the natural language of disease; it assumes shape and value only within the questions posed by medical investigation. There is nothing, therefore, to prevent it being solicited and almost fabricated by medical investigation. It is no longer that which is spontaneously stated by the disease itself; it is the meeting point of the gestures of research and the sick organism. This explains why Corvisart was able, without any major theoretical problem, to reactivate Auenbrugger's relatively old and completely forgotten discovery. This discovery was based on well-founded pathological knowledge: the diminution of the volume of air contained by the thoracic cavity in many pulmonary affections. It was also explained by a datum of simple experience: the degree of dullness of the sound produced when a barrel is struck indicates the degree to which it is filled. Lastly, it was justified by experimentation on corpses: 'If in a corpse the sound cavity of the thorax is filled with liquid by means of injection, then the sound, on the side of the chest that has been filled, becomes deadened up to the height reached by the injected liquid' [36].

It was natural that clinical medicine at the end of the eighteenth century should ignore a technique that made a sign appear artificially where there had been no symptom, and solicited a response when the disease itself did not speak: a clinic as expectant in its reading as in its therapeutics. But as soon as pathological anatomy compels the clinic to question the body in its organic density, and to bring to the surface what was given only in deep layers, the idea of a technical artifice capable of surprising a lesion becomes once again a scientifically based idea. The return to Auenbrugger can be explained by the same reorganization of structures as the return to Morgagni. Sounding by percussion is not justified if the disease is composed only of a web of symptoms; it becomes necessary if the patient is hardly more than an injected corpse, a half-filled barrel.

To establish these signs, artificial or natural, is to project upon the living body a whole network of anatomo-pathological mappings: to draw the dotted outline of the future autopsy. The problem, then, is to bring to the surface that which is layered in depth; semiology will no longer be a *reading*, but the set of techniques that make it possible to constitute a *projective pathological anatomy*. The clinician's gaze was directed upon a succession and upon an area of pathological events; it had to be both synchronic and diachronic, but in any case it was placed under temporal obedience; it *analysed a series*. The anatomo-clinician's gaze has *to map a volume*; it deals with the complexity of spatial data which for the first time in medicine are three-dimensional. Whereas clinical experience implied the constitution of *a mixed web of the visible and the readable*, the new semiology requires a sort of *sensorial triangulation* in which various atlases, hitherto excluded from medical techniques, must collaborate: the ear and touch are added to sight.

For thousands of years, after all, doctors had tested patients' urine. Later, they began to touch, tap, listen. Was this the result of the raising of moral prohibitions by the Enlightenment? If such was the case, it would be difficult to understand why, under the Empire, Corvisart should have reintroduced percussion, or why, under the Restoration, Laënnec should have put his ear, for the first time, to women's breasts. The moral obstacle was experienced only when the epistemological need had emerged; scientific necessity revealed the prohibition for what it was: Knowledge invents the Secret. Zimmermann, in order to discover the force of the circulation, had expressed a wish that 'doctors should be free to make their observations in this respect by placing their hands directly on the heart'; but he added that 'our delicate morals prevent us from doing so, especially in the case of women' [37]. In 1811, Double criticized this 'false modesty', this 'excessive restraint'; not that he believed that such a practice should be carried out without any reserve whatsoever: 'this exploration, which is carried out very precisely above the chemise, may take place with all possible decency' [38]. The moral screen, the need for which was recognized, was to become a technical mediation. The *libido sciendi*, strengthened by the prohibition that it had aroused and discovered, circumvents it by making it more imperious; it provides it with scientific and social justifications, inscribing it within necessity in order to pretend the more easily to efface it from the ethical, and to build upon it the structure that traverses it and maintains it. It is no longer shame that prevents contact, but dirt and poverty; not the innocence, but the disgrace, of the body. Auscultation is not only direct, but 'inconvenient for both doctor and patient; only disgust makes it more or less impracticable in hospitals; it is scarcely mentionable in the case of most women, and in the case of some women, the size of the breasts is a physical obstacle to its practice'. The stethoscope is the measure of a prohibition transformed into disgust, and a material obstacle:

In 1816, I was consulted by a young person who presented symptoms of heart disease, and in the case of whom the application of
the hand and percussion yielded poor results on account of her
plumpness of figure. Since the age and sex of the patient forbade
me the kind of examination of which I have just spoken (the application of the ear to the precordial region), I happened to recall
a well-known acoustical phenomenon: if one places one's ear at
the end of a beam, one can hear very distinctly a pin dropped on
to the other end [39].

The stethoscope, solidified distance, transmits profound and invisible
events along a semi-tactile, semi-auditory axis. Instrumental mediation outside the body authorizes a withdrawal that measures the
moral distance involved; the prohibition of physical contact makes it
possible to fix the virtual image of what is occurring well below the
visible area. For the hidden, the distance of shame is a projection
screen. What one *cannot* see is shown in the distance from what
one *must not* see.

Thus armed, the medical gaze embraces more than is said by
the word 'gaze' alone. It contains within a single structure different
sensorial fields. The sight/touch/hearing trinity defines a perceptual
configuration in which the inaccessible illness is tracked down by
markers, gauged in depth, drawn to the surface, and projected
virtually on the dispersed organs of the corpse. The 'glance' has
become a complex organization with a view to a spatial assignation
of the invisible. Each sense organ receives a partial instrumental
function. And the eye certainly does not have the most important
function; what can sight cover other than 'the tissue of the skin
and the beginning of the membranes'? Through touch we can locate
visceral tumours, scirrhous masses, swellings of the ovary, and dila-
tions of the heart; while with the ear we can perceive 'the crepitation of fragments of bone, the rumbling of aneurism, the more or
less clear sounds of the thorax and the abdomen when sounded' [40].
The medical gaze is now endowed with a plurisensorial structure.
A gaze that touches, hears, and, moreover, not by essence or necessity, sees.

Let me quote a historian of medicine: 'As soon as one used the
ear or the finger to recognize on the living body what was revealed
on the corpse by dissection, the description of diseases, and therefore therapeutics took a quite new direction' [41].

But we must not lose sight of the essential. The tactile and auditory dimensions were not simply added to the domain of vision. The sensorial triangulation indispensable to anatomo-clinical perception remains under the dominant sign of the visible: first, because this multi-sensorial perception is merely a way of anticipating the triumph of the gaze that is represented by the autopsy; and ear and hand are merely temporary, substitute organs until such time as death brings to truth the luminous presence of the visible; it is a question of a mapping in life, that is, in *night*, in order to indicate how things would be in the white brightness of death. And above all, the alterations discovered by anatomy concern 'the shape, the size, the position, and the direction' of organs or of their tissues [42]: that is, spatial data that belong by right of origin to the gaze. When Laënnec speaks of alterations of structure, it is never a question of what is beyond the visible, or even of what would be perceptible to a delicate touch, but of solutions of continuity, accumulations of liquids, abnormal increases, or inflammations indicated by the swelling and redness of the tissue [43]. In any case, the absolute limit and the depth of perceptual exploration are always outlined by the clear plane of an at least potential visibility. 'They are painting a picture', says Bichat of the anatomists, 'rather than learning things. They must see rather than meditate' [44]. When Corvisart hears a heart that functions badly or Laënnec a voice that trembles, what they see with that gaze that secretly haunts their hearing and, beyond it, animates it, is a hypertrophy, a discharge.

Thus, from the discovery of pathological anatomy, the medical gaze is duplicated: there is a local, circumscribed gaze, the borderline gaze of touch and hearing, which covers only one of the sensorial fields, and which operates on little more than the visible surfaces. But there is also an absolute, absolutely integrating gaze that dominates and founds all perceptual experiences. It is this gaze that structures into a sovereign unity that which belongs to a lower level of the eye, the ear, and the sense of touch. When the doctor observes, with all his senses open, another eye is directed upon the fundamental visibility of things, and, through the transparent datum of life with which the particular senses are forced to work, he addresses himself fairly and squarely to the bright solidity of death.

The structure, at once perceptual and epistemological, that commands clinical anatomy, and all medicine that derives from it, is that of *invisible visibility*. Truth, which, by right of nature, is made for the eye, is taken from her, but at once surreptitiously revealed by that which tries to evade it. Knowledge *develops* in accordance with a whole interplay of *envelopes*; the hidden element takes on the form and rhythm of the hidden content, which means that, like a *veil*, it is *transparent* [45]: the aim of the anatomists 'is attained when the opaque envelopes that cover our parts are no more for their practised eyes than a transparent veil revealing the whole and the relations between the parts' [46]. The individual senses lie in wait through these envelopes, try to circumvent them or lift them up; their lively curiosity invents innumerable means, including even making shameless use of the sense of shame (witness the stethoscope). But the absolute eye of knowledge has already confiscated, and re-absorbed into its geometry of lines, surfaces, and volumes, raucous or shrill voices, whistlings, palpitations, rough, tender skin, cries—a suzerainty of the visible, and one all the more imperious in that it associates with it power and death. That which hides and envelops, the curtain of night over truth, is, paradoxically, life; and death, on the contrary, opens up to the light of day the black coffer of the body: obscure life, limpid death, the oldest imaginary values of the Western world are crossed here in a strange misconstruction that is the very meaning of pathological anatomy if one agrees to treat it as a fact of civilization of the same order as —and why not?—the transformation from an incinerating to an in-huming culture. Nineteenth-century medicine was haunted by that absolute eye that cadaverizes life and rediscovers in the corpse the frail, broken nervure of life.

In former times, doctors communicated with death by means of the great myth of immortality or at least of the gradually receding limits of existence [47]. Now, these men who watch over men's lives communicate with their death in the fine, rigorous form of the gaze.

However, this projection of illness onto the plane of absolute visibility gives medical experience an opaque base beyond which it can no longer go. That which is not on the scale of the gaze falls outside the domain of possible knowledge. Hence the rejection of a number of scientific techniques that were nonetheless used by doctors in earlier years. Bichat even refused to use the microscope: 'when one looks into darkness everyone sees in his own way' [48]. The only type of visibility recognized by pathological anatomy is that defined by everyday vision: a *de jure* visibility that envelops in temporary invisibility an opaque transparency, and not (as in microscopic investigation) a *de natura* invisibility that is breached for a time by an artificially multiplied technique of the gaze. In a way that seems strange to us, but that was structurally necessary, the analysis of pathological tissues dispensed, over a period of several years, with even the most ancient instruments of optics.

Still more significant is the rejection of chemistry. Analysis, as practised by Lavoisier, served as an epistemological model for the new anatomy [49], but it did not function as a technical extension of his gaze. In eighteenth-century medicine there was no dearth of experimental ideas; when one wanted to know what inflammatory fever consisted of, one carried out blood analyses: the average weight of the coagulated mass was compared with that of 'the lymph that separates from it'; distillations were made, and measurements were taken of the masses of fixed and volatile salt, oil, and earth to be found in a patient and in a healthy subject [50]. At the beginning of the nineteenth century, this experimental apparatus disappeared, and the only remaining technical problem was to know whether the opening up of the corpse of the patient affected by inflammatory fever would or would not reveal visible alterations. 'In order to characterize a morbid lesion,' Laënnec explains, 'it is usually enough to describe its physical or perceptible characteristics, and to indicate the course it takes in its development and in its terminations'; at most, one has time to use certain 'chemical reactions' only if they are very simple and intended to 'reveal certain physical characteristics': thus one may heat a liver, or pour an acid onto a degenerescence of which one is not sure whether it is fatty or albuminous [51].

Alone, the gaze dominates the entire field of possible knowledge; the intervention of techniques presenting problems of measurement, substance, or composition at the level of invisible structures is rejected. Analysis is not carried out in the sense of an indefinite descent towards the finest configurations, ultimately to those of the inorganic; in that direction, it soon comes up against the absolute limit laid down for it by the gaze, and from there, taking the perpendicular, it slides sideways towards the differentiation of individual qualities. On the line on which the visible is ready to be resolved into the invisible, on that crest of its disappearance, singularities come into play. A discourse on the individual is once more possible, or, rather, necessary, because it is the only way in which the gaze can avoid renouncing itself, effacing itself in the figures of experience, in which it would be disarmed. The principle of visibility has its correlative in the differential reading of cases.

The process of such a reading is very different from clinical experience in its earliest form. The analytical method would consider the case only in its function as a semantic support; the forms of coexistence or of the series in which it was caught up made it possible to annul in it whatever was accidental or variable; its legible structure appeared only in the neutralization of what was not essential. The clinic was a science of cases to the extent that it proceeded initially to the diminution of individualities. In the anatomic method, individual perception is given at the term of a spatial quadrilateral of which it constitutes the finest, most differentiated structure, and, paradoxically, the one most open to the accidental, while at the same time being the most explanatory. Laënnec observes a woman who presents the typical symptoms of a heart affection: pale, puffy face, purple lips, infiltrated lower extremities, short, accelerated, panting breathing, coughing fits, inability to lie down. The opening up of the corpse shows pulmonary phthisis with concretionary cavities, and tubercles yellowish at the centre, grey and transparent around the circumference. The heart was in an almost natural state (except for the right auricle, which was very distended). But the left lung adhered to the pleura by a cellulous wrinkle, and was covered with irregular, convergent stripes in that area; the top of the lung presented fairly broad, crossed strips [52]. This particular kind of tuberculous lesion accounted for the impeded, rather suffocated, breathing and the circulatory alterations, which gave the clinical picture of a distinctly cardiac appearance. For the first time, the anatomo-clinical method integrates into the structure of the illness the constant possibility of an individual modulation. This possibility existed, of course, in earlier medicine: but it was conceived only in the abstract form of the subject's temperament, or of influences due to the environment, or of therapeutic interventions intended to alter a pathological type from the outside. In anatomical perception, the disease is given only with a certain 'blurring'; it has, from the outset, a latitude of insertion, direction, intensity, and acceleration that forms its individual figure. This figure is not a deviation added to the pathological deviation; the disease is itself a perpetual deviation within its essentially deviant nature. Only individual illnesses exist:

not because the individual reacts upon his own illness, but because the action of the illness rightly unfolds in the form of individuality.

Hence the new turn given to medical language. It is no longer a question, by means of a bi-univocal placing in correspondence, of promoting the visible to the legible, and of turning it into the significative by means of the universality of a codified language; but, on the contrary, of opening words to a certain qualitative, ever more concrete, more individualized, more modelled refinement; the importance of colour, consistency, texture, a preference for metaphor rather than measurement (as big as . . . , of the size of a . . .); an appreciation of the ease or difficulty to be found in simple operations (tearing, crushing, pressing); the value of inter-sensorial qualities (smooth, greasy, bumpy); empirical comparisons and references to the everyday or normal (deeper than in the natural state, an intermediate sensation 'between that of a damp bladder half-filled with air that one squeezes between the fingers and the natural crepitation of a healthy pulmonary tissue') [53]. It is no longer a question of correlating a perceptual sector and a semantic element, but of bending language back entirely towards that region in which the perceived, in its singularity, runs the risk of eluding the form of the word and of becoming finally imperceptible because incapable of being said. To *discover*, therefore, will no longer be to *read* an essential coherence beneath a state of disorder, but to push a little farther back the foamy line of language, to make it encroach upon that sandy region that is still open to the clarity of perception but is already no longer so to everyday speech —to introduce language into that penumbra where the gaze is bereft of words. An arduous, delicate work; a work that *reveals*, as Laënnec revealed distinctly, outside the confused mass of scirrhi, the first cirrhotic liver in the history of medical perception. The extraordinary formal beauty of the text links, in a single movement, the internal work of a language in pursuit of perception with all the strength of its stylistic originality, and the conquest of a hitherto unperceived pathological individuality:

> The liver, reduced to a third of its volume, was, as it were, hidden in the region that it occupies; its external surface, slightly mammil-lated and emptied, was a yellowish grey in colour; when cut, it seemed to be made up entirely of a mass of small seeds, round or oval in shape, varying in size from a millet seed to a hemp seed. These seeds, which can be easily separated, left almost no gap

between them in which one might be able to make out some remaining part of the real tissue of the liver; they were fawn or
reddish-yellow in colour, verging in parts on the greenish; their
fairly moist, opaque tissue was slack, rather than soft, to the touch,
and when one squeezed the grains between one's fingers only a
small part was crushed, the rest feeling like a piece of soft leather
[54].

The figure of the visible invisible organizes anatomo-pathological
perception. But, as one sees, in accordance with a reversible structure. It is a question of the *visible* that the living individuality, the
intersection of symptoms, the organic depth, in fact, and for a time,
render invisible, before the sovereign resumption of the anatomical
gaze. But it is as much a question of this *invisible* of the individual
modulations, whose extrication seemed impossible even to a clinician like Cabanis [55], and which the effort of an incisive, patient,
eroding language offers at last to common light what is *visible* for
all. Language and death have operated at every level of this experience, and in accordance with its whole density, only to offer at last
to scientific perception what, for it, had remained for so long the
visible invisible—the forbidden, imminent secret: the knowledge
of the individual.

The individual is not the initial, most acute form in which life
is presented. It was given at last to knowledge only at the end of a
long movement of spatialization whose decisive instruments were a
certain use of language and a difficult conceptualization of death.
Bergson is strictly in error when he seeks in time and against space,
in a silent grasp of the internal, in a mad ride towards immortality,
the conditions with which it is possible to conceive of the living
individuality. Bichat, a century earlier, gave a more severe lesson.
The old Aristotelian law, which prohibited the application of scientific discourse to the individual, was lifted when, in language,
death found the locus of its concept: space then opened up to the
gaze the differentiated form of the individual.
According to the order of historical correspondences, this introduction of death into knowledge goes very far: the late eighteenth century rediscovered a theme that had lain in obscurity since
the Renaissance. To see death in life, immobility in its change,
skeletal, fixed space beneath its smile, and, at the end of its time,
the beginning of a reversed time swarming with innumerable lives,

is the structure of a Baroque experience whose re-appearance was attested by the previous century four hundred years after the frescoes of Campo Santo. Is not Bichat, in fact, the contemporary of the man who suddenly, in the most discursive of languages, introduced eroticism and its most inevitable point, death? Once more, knowledge and eroticism denounce, in this coincidence, their profound kinship. Throughout the latter years of the eighteenth century, this kinship opened up death to the task, to the infinitely repeated attempts of language. The nineteenth century will speak obstinately of death: the savage, castrated death of Goya, the visible, muscular, sculptural death offered by Géricault, the voluptuous death by fire in Delacroix, the Lamartinian death of aquatic effusions, Baudelaire's death. To know life is given only to that derisory, reductive, and already infernal knowledge that only wishes it dead. The Gaze that envelops, caresses, details, atomizes the most individual flesh and enumerates its secret bites is that fixed, attentive, rather dilated gaze which, from the height of death, has already condemned life.

But the perception of death in life does not have the same function in the nineteenth century as at the Renaissance. Then it carried with it reductive significations: differences of fate, fortune, conditions were effaced by its universal gesture; it drew each irrevocably to all; the dances of skeletons depicted, on the underside of life, a sort of egalitarian saturnalia; death unfailingly compensated for fortune. Now, on the contrary, it is constitutive of singularity; it is in that perception of death that the individual finds himself, escaping from a monotonous, average life; in the slow, half-subterranean, but already visible approach of death, the dull, common life becomes an individuality at last; a black border isolates it and gives it the style of its own truth. Hence the importance of the Morbid. The *macabre* implied a homogeneous perception of death, once its threshold had been crossed. The *morbid* authorizes a subtle perception of the way in which life finds in death its most differentiated figure. The morbid is the *rarefied* form of life, exhausted, working itself into the void of death; but also in another sense, that in death it takes on its peculiar volume, irreducible to conformities and customs, to received necessities; a *singular* volume defined by its absolute rarity. The privilege of the consumptive: in earlier times, one contracted leprosy against a background of great waves of collective punishment; in the nineteenth century, a man, in becoming tubercular, in the fever that hastens things and betrays them, fulfills his incommunicable secret. That is why chest diseases are of exactly the same nature as diseases of love: they are the Passion, a life to which death gives a face that cannot be exchanged. Death left its old tragic heaven and became the lyrical core of man: his invisible truth, his visible secret.

NOTES

[1] Roederer and Wagler, *Tractatus de morbo mucoso* (Göttingen, 1783).
[2] X. Bichat, *Anatomie générale*, avant-propos, vol. I, p. lxxxv.
[3] X. Bichat, *Traité des membranes*, ed. Magendie, pp. 122-3.
[4] *Ibid.*, p. 101.
[5] Bichat, *Anatomie générale*, vol. I, avant-propos, p. xci.
[6] *Ibid.*, p. xcii.
[7] Corvisart, *Essai sur les maladies et les lésions organiques du coeur et des gros vaisseaux.*
[8] G.-L. Bayle, *Recherches sur la phtisie pulmonaire*, pp. 13-14.
[9] Cf. above, p. 132.
[10] Article 'Anatomie pathologique', in *Bulletin de l'École de Médecine de Paris*, Year XIII, first issue, pp. 16-18.
[11] Bayle, *op. cit.*, p. 12.
[12] *Ibid.*, pp. 423-4.
[13] F. Lallemand, *Recherches anatomo-pathologiques sur l'encéphale et ses dépendances*, I, pp. 98-9.
[14] Bichat, *Anatomie générale*, vol. IV, p. 591.
[15] *Ibid.*, I, avant-propos, p. vii.
[16] F.-J. Broussais, *Histoire des phlegmasies chroniques* (Paris, 1808, vol. I, pp. 54-5).
[17] Cf., for example, Pinel, *Nosographie philosophique*, introduction, p. xi, or C.-L. Dumas, *Recueil de discours prononcés à la Faculté de Médecine de Montpellier* (Montpellier, 1820, pp. 22-3).
[18] Buffon, 'Histoire naturelle', *Oeuvres complètes* (Paris, 1848, vol. III, p. 311).
[19] Corvisart, *op. cit.*, pp. 636-7.
[20] *Ibid.*, p. 636, n. 1.
[21] Cf. *Bulletin des sciences médicales*, vol. V, 1810.
[22] J. Cruveilhier, *Anatomie pathologique* (Paris, 1816, vol. I, pp. 75-6).
[23] R. Laënnec, article on 'Dégénération', *Dictionnaire des Sciences médicales*, 1814, vol. VIII, pp. 201-7.
[24] R. Laënnec, Introduction and first chapter of *Traité inédit d'anatomie pathologique*, p. 52.
[25] Dupuytren, *Dissertation inaugurale sur quelques points d'anatomie*, Paris, Year XII, p. 21.
[26] Lallemand, *op. cit.*, I, pp. 88-9.

[27] Bayle, *op. cit.*, p. 5.
[28] Corvisart, *op. cit.*, discours, préliminaire, p. xvii.
[29] Cf. above, p. 93.
[30] Bayle, *op. cit.*, pp. 5–14.
[31] Laënnec, *Traité de l'auscultation médiate* (Paris, 1819, vol. I).
[32] A.-F. Chomel, *Éléments de pathologie générale* (Paris, 1817, pp. 522–3).
[33] Menuret, *Nouveau traité du pouls* (Amsterdam, 1768, pp. ix–x).
[34] Bordeu, *Recherches sur le pouls* (Paris, 1771, vol. I, pp. 30–1).
[35] Corvisart, *op. cit.*, pp. 397–8.
[36] Auenbrugger, *Nouvelle méthode pour reconnaître les maladies internes de la poitrine* (trans. Corvisart, Paris, 1808, p. 70).
[37] G. Zimmermann, *Traité de l'expérience en médicine* (Fr. trans., Paris, 1774, II, p. 8).
[38] F.-J. Double, *Séméiologie générale.*
[39] Laënnec, *op. cit.*, vol. I, pp. 7–8.
[40] A.-F. Chomel, *Éléments de pathologie générale* (Paris, 1817, pp. 30–1).
[41] Ch. Daremberg, *Histoire des sciences médicales* (Paris, 1870, II, p. 1066).
[42] X. Bichat, essay on Desault, *Oeuvres chirurgicales de Desault*, 1798, I, pp. 10 and 11.
[43] Laënnec, *Dictionnaire des sciences médicales*, vol. II, article on 'Anatomie pathologique', p. 52.
[44] Bichat, *op. cit.*, I, p. 11.
[45] This structure does not date from the beginning of the nineteenth century, far from it; in its general outline, it dominated the forms of knowledge and eroticism in Europe from the mid-eighteenth century onwards, and it prevailed until the end of the nineteenth century. I shall try to study it in a subsequent work.
[46] Bichat, *op. cit.*, I, p. 11.
[47] Cf., still at the end of the eighteenth century, a text like that of Hufeland, *Makrobiotik oder der Kunst das Leben zu verlängern* (Jena, 1796).
[48] X. Bichat, *Traité des membranes* (Paris, Year VIII, p. 321).
[49] Cf. above, Chapter 8.
[50] Experiments carried out by Langrish and Tabor, cited by Sauvages, *Nosologie méthodique*, vol. II, pp. 331–3.
[51] R. Laënnec, Introduction and first chapter, *Traité inédit d'anatomie pathologique* (Published by V. Cornil, Paris, 1884, pp. 16–17).
[52] R. Laënnec, *De l'auscultation médiate*, vol. I, pp. 72–6.
[53] *Ibid.*, p. 249.
[54] *Ibid.*, p. 368.
[55] Cf. above.

10 · Crisis in Fevers

In this chapter we shall examine the final process by which anatomo-clinical perception finds the form of its equilibrium. If we allowed ourselves to become involved in the detail of events, it would be a long chapter indeed: for almost twenty-five years (from 1808, which saw the appearance of the *Histoire des phlegmasies chroniques*, to 1832, when their place was largely taken over by discussions on cholera), the theory of essential fevers and Broussais's critique of it occupied a considerable area in medical research, more considerable, indeed, than was warranted by a problem that could be settled so quickly at the level of observation; but the sheer quantity of the polemics, the difficulty of reaching an understanding when one was in agreement as to the facts, the wide use of arguments that had little or nothing to do with pathology indicate an essential confrontation, the last (and the most violent, most complex) of the conflicts between two incompatible types of medical experience.

The method constituted by Bichat and his earliest followers left open two series of problems.

The first concerned the very being of disease and its relation to lesional phenomena. When one observes a serous discharge, a degenerated liver, a perforated lung, is what one sees pleurisy, cirrhosis, or phthisis themselves, in all their pathological depth? Is the lesion the original, tri-dimensional form of the disease, which is thus spatial in nature—or must it be situated beyond, in the region of proximate causes, or immediately prior to it, as the first visible manifestation of a process that remains hidden? It is clear enough—after the event—what reply is prescribed by the logic of anatomo-clinical perception: for those who practised this perception for the first time in the history of medicine, things were not so clear. N.-A. Petit, who based his whole conception of entero-mesenteric fever on observations of pathological anatomy, did not think that in the intestinal lesions accompanying certain so-called adynamic or ataxic fevers he had discovered the very essence of the disease, or its ultimate truth; for him, these lesions were merely the 'seat' of the disease, and this geographical determination was of less importance for medical knowledge than 'the general set of symptoms that distinguish one disease from another and reveal their true character': so much so that therapy goes astray when it seeks to treat intestinal lesions instead of following the indications of symptomatology, which prescribes tonics [1]. The 'seat' is merely the spatial insertion of the disease; it is the other morbid manifestations that designate its essence. This essence remains the great prerequisite that links cause to symptom, thus throwing the lesion back into the domain of the accidental; the tissual or organic attack marks only the approach point of the disease, the region from which its colonizing enterprise will develop:

> Between the hepatization of the lung and the causes that bring it about, something occurs that eludes us; it is the same with all the lesions encountered on opening up a corpse; far from being the first cause of all the phenomena observed, they are themselves the effect of a particular disorder in the secret action of our organs; and this ultimate action eludes all our methods of investigation [2].

As pathological anatomy becomes more accurate in situating the seat of the disease, it would seem that the disease itself withdraws ever more deeply into the intimacy of an inaccessible process.

There is another series of questions: Do all diseases have their lesional correlative? Is the possibility of assigning a seat to them a general principle of pathology, or does it concern only a very special group of morbid phenomena? And if the latter, is it possible to begin the study of diseases with a nosographical type of classification (organic disorders/non-organic disorders) before entering the domain of pathological anatomy? Bichat had made room for non-lesional diseases—but he did little more than treat them by preterition: 'take away certain kinds of fevers and nervous affections and everything is almost in the domain of this science' (pathological anatomy) [3]. Laënnec accepts from the outset the division of diseases into

> two great classes: those that are accompanied by a lesion present in one or several organs: for several years these have been known as organic diseases; and those that leave in no part of the body an alteration that is constant and to which an origin may be attributed: these are what are commonly called nervous diseases [4].

At the time Laënnec wrote this text (1812), he had not yet taken up a definitive position in relation to the fevers: he was still close to the localizers from whom he was soon to break away. At the same time, Bayle distinguished the *organic*, not from the *nervous*, but from the *vital*, as opposed to organic lesions, vices of solids (tumefactions, for example), vital disorders, 'alterations of vital properties or functions' (pain, heat, acceleration of the pulse); and one may be superimposed upon another, as in phthisis [5]. It was this classification that Cruveilhier was soon to take up in a rather more complex form: organic, simple, and mechanical lesions (fractures), originally organic and secondarily vital lesions (haemorrhages); originally vital affections combined with organic lesions, whether deep (chronic phlegmasias) or superficial (acute phlegmasias); lastly, vital diseases involving no lesion (neuroses and fevers) [6].

However much it was said that the whole domain of nosology remained under the control of pathological anatomy, and that a vital disease could be proved to be so only negatively, and by a failure to discover any lesions, it was nonetheless the case that by this same detour a form of classificatory analysis was rediscovered. Its species—and not its seat or its cause—determined the nature of a disease; and the very fact of having or not having a localizable site was prescribed by the prior forms of this determination. The lesion was not the disease, but merely the first of the manifestations by which this generic character appeared, which opposed it to affections possessing no support. Paradoxically, the concern of the anatomo-pathologists revitalized the classificatory idea. It is this that gives Pinel's work its meaning and its curious prestige. His thinking was developed at Montpellier and in Paris, in the tradition of Sauvages and under the more recent influence of Cullen, and was classificatory in structure; but it had the good and ill fortune at once to develop at a time when first the clinical theme, then the anatomo-clinical method were depriving nosology of its real content, but not without producing effects—though temporary ones—of mutual reinforcement: we have seen how the idea of class was correlative with a certain neutral observation of symptoms [7], how clinical decipherment involved a reading of essences [8]; we are now seeing how pathological anatomy was ordered, quite spontaneously, in accordance with a certain form of nosography. Pinel's entire work owes its strength to each of these reinforcements: his method only secondarily requires the clinic or the anatomy of lesions; basically, it is the organization, in accordance with a real, but abstract, coherence, of the temporary structures by which the clinical gaze or the anatomo-pathological perception sought their support or momentary equilibrium in the already existent nosology. None of the doctors of the old school was better disposed to the new forms of medical experience; he readily took on teaching duties and carried out autopsies without too much reluctance; but he perceived only the effects of recurrence, following, in the birth of the new structures, only the outlines that they derived from the old [9]: so much so that nosology was constantly being confirmed, and the new experience contained in advance. Bichat was perhaps alone in understanding from the outset the incompatibility of its method with that of the nosographers: 'We discover the procedures of nature as best we can. . . . Let us not attach exaggerated importance to this or that classification': none will ever give us 'a precise picture of the progress of nature' [10]. Laënnec, on the other hand, found no difficulty in enveloping the anatomo-clinical experience in the space of the nosological division: opening up corpses and finding lesions was to reveal 'the most fixed, most positive, and least variable elements in local diseases'; it was therefore to isolate 'that which must characterize or specify them'; by providing it with more certain criteria, it was, in the last resort, to serve the cause of nosology [11]. It was in this spirit that the Société d'Émulation, which grouped together the younger generation and faithfully represented the new school, asked at the *concours* of 1809 the famous question: 'What diseases may be specially regarded as organic?' [12] What was in question was certainly the notion of essential fever and its non-organicity—a notion to which Pinel had remained attached—but on this precise point the problem presented was still one of species and class. Pinel was discussed, but his medicine was given no radical revaluation.

This was not done until 1816, when Broussais published his *Examen de la Doctrine généralement admise*, in which he expressed in a more radical form the criticisms that he had already formulated eight years before in his *Histoire des phlegmasies chroniques*. In an unexpected way, this explicitly physiological medicine, this easy, loose theory of sympathies, the general use of the concept of irritation, and hence the return to a certain pathological monism closely related to that of Brown were needed if pathological anatomy was to be really freed from the tutelage of the nosographers and the problem of morbid essences cease to duplicate the perceptual analysis of organic lesions. In time, it would be forgotten that the structure of anatomo-clinical experience attained equilibrium only thanks to Broussais; only his frenzied attacks on Pinel would be remembered—Pinel, whose impalpable control Laënnec, on the other hand, supported so well; only the intemperate physiologist and his hasty generalizations would be remembered. And recently, the good Mondor found beneath the benignity of his pen fresh adolescent insults to hurl at Broussais's departed shade [13]. The imprudent man had not read the texts, or understood very much.

At the end of the eighteenth and the beginning of the nineteenth centuries, neuroses, and essential fevers were fairly generally regarded as diseases without organic lesion. Diseases of the mind and nerves had, thanks to Pinel, been given a sufficiently special status for their history, at least until A.-L. Bayle's discovery of 1821-1824, to be quite distinct from discussions concerning the organicity of diseases. For over fifteen years, fevers, on the other hand, were at the very centre of the problem.

First, let us go over some of the general lines of the eighteenth-century concept of fever. In the first instance, the term was understood to mean a finalized reaction of the organism defending itself against a pathogenic attack or substance; the fever that appears in the course of the disease goes in the opposite direction and tries to stem the current; it is a sign not of the disease, but of the resistance to the disease, 'an affection of life striving to break away from death' [14]. In the strict sense of the term, it has, therefore, a salutary value: it shows that the organism 'morbiferam aliquam materiam sive praeoccupare sive removere intendit' [15]. Fever is

an excretory movement, purificatory in intention; and Stahl recalls
an etymology: *februare* is to expel ritually from a house the shades
of the dead [16].

Against this background of finality, it was easy enough to
analyse the movement of the fever and its mechanism. The succession of the symptoms indicates its different phases: the shiver and
the first impression of coldness indicate a peripheric spasm and a
rarefaction of the blood in the capillaries close to the skin. The
increased pulse rate indicates that the heart is reacting by making
as much blood as possible flow out towards the limbs: the heat
shows that the blood is in fact circulating more rapidly, and that
all the functions are thereby accelerated; the motor forces decrease
proportionally: hence the impression of languor and the atony of
the muscles. Finally, sweating indicates that this feverish reaction
is succeeding in expelling the morbific substance; but when this
succeeds in reforming itself in time, one suffers from intermittent
fevers [17].

This simple interpretation, which linked in so evident a fashion
the manifest symptoms to their organic correlatives, was of triple
importance in the history of medicine. First, in its general form,
the analysis of fever corresponds exactly with the mechanism of
local inflammations; in each case there is a fixation of blood, a contraction causing a more or less extended stasis, followed by an
effort of the system to re-establish the circulation, and as a result
a violent movement of the blood; it will be seen that 'red globules
begin to pass into the lymphatic arteries', which causes, in a local
form, the injection of the conjunctive, for example, or, in a general
form, heat and movement throughout the whole organism; if the
movement increases, the most tenuous parts of the blood separate
from the heavier, which remain in the capillaries, where 'the lymph
will be converted into a sort of jelly': hence the suppurations that
occur in the respiratory or intestinal system in cases of generalized
inflammation, or abscesses in cases of local fever [18].

But if there is a functional identity between inflammation and
fever it is because the circulatory system is the essential element
of the process. There is a double shift in the normal functions: first
a slowing down, then an exaggeration; first an irritating phenomenon, then a phenomenon of irritation. 'All these phenomena must
be deduced from the irritability of the heart and arteries increased
and stimulated, and finally from the action of some stimulus and
from the resistance of life thus irritated to the detrimental stimulus

[19]. Thus fever, whose intrinsic mechanism may be either general or local, finds in the blood the organic, isolable support that may render it local or general, or first local, then general. By this diffused irritation of the blood system, a fever may always be the general symptom of a disease that remains local throughout its development: without anything being modified as to its mode of action, it may be either essential or sympathetic. In such a schema, the problem of the existence of essential fevers without assignable lesions could not arise: whatever its form, its starting point, or its surface of manifestation, fever always had the same type of organic support.

Finally, the phenomenon of heat is far from constituting the essence of the febrile movement; it is no more than its most superficial and most transitory culmination, whereas the movement of the blood, the impurities that it absorbs or expels, the obstructions or exudations that occur indicate what the essential nature of fever is. Grimaud warns against physical instruments that

> can certainly indicate no more than the degrees of intensity of heat; and these differences are the least important in practice; . . . the doctor must apply himself, above all, to distinguishing in feverish heat qualities that may be perceived only by a highly practised touch, and which elude whatever means physics may offer. There is, for example, that acrid, irritating quality of feverish heat

that gives the same impression as 'smoke in the eyes', and that points to a putrid fever [20]. Below the homogeneous phenomenon of heat, fever has, therefore, its own qualities, a sort of substantial, differentiated solidity that makes it possible to divide it up according to specific forms. It is possible, accordingly, to pass naturally and unproblematically from *fever* to *fevers*. The shift in meaning and epistemological level, which seems so striking to us [21], between the designation of a common symptom and the determination of specific diseases, cannot be perceived by eighteenth-century medicine, given the form of analysis by which it deciphered the febrile mechanism.

In the name, therefore, of a highly homogeneous, coherent conception of 'fever', the eighteenth century was to adopt a considerable number of 'fevers'. Stoll recognized twelve, to which he added 'new and unknown' fevers. They were specified either by the circulatory mechanism that explained them (the inflammatory fever analysed by J.-P. Franck, and traditionally designated as synochus), or according to the organs in which the inflammation occurred (Baglivi's mesenteric fever), or according to the quality of the excretions caused by it (the putrid fever referred to by Haller, Tissot, and Stoll), or according to the variety of forms that it assumed and its possible evolution (Selle's malign fever or ataxic fever). To our clouded eyes, this network became confused only when the medical gaze changed structure.

The first meeting between anatomy and the symptomatic analysis of fevers took place well before Bichat, and well before Prost's first observations. It was a purely negative meeting, since the anatomical method abandoned its rights and ceased to assign a seat to certain feverish diseases. In the forty-ninth letter of his *Treatise*, Morgagni said that on opening up patients who had died of violent fevers he had found 'vix quidquam . . . quod earum gravitati aut impetui responderet; usque adeo id saepe latet per quod faber interficiunt' [22]. An analysis of fevers based only on their symptoms, with no attempt at localization, became not only possible but necessary: in order to provide the different forms of fever with a structure, organic volume had to be replaced by a space of division occupied only by signs and what they signify.

The re-ordering brought about by Pinel was not only in line with his own method of nosological decipherment; it was contained exactly within the structure defined by this first form of pathological anatomy: fevers without lesions are essential fevers; those with local lesions are sympathetic fevers. These idiopathic forms, which are characterized by their external manifestations, reveal

common properties such as the suspension of the appetite and digestion, the alteration of the circulation, the interruption of certain secretions, the prevention of sleep, the excitation or diminution of the activity of hearing, the alteration, or even the suspension, of certain functions of the senses, and the hindering, each in its own way, of muscular movement [23].

But the diversity of the symptoms also makes possible the reading of different species: an inflammatory or angiotonic form 'marked on the outside by signs of irritation or tension of the blood vessels' (it is frequent at puberty, at the onset of pregnancy, and after alcoholic excesses); a 'meningo-gastric' form with nervous symptoms, but also with other, more primitive ones that appear 'to correspond with the epigastric region' and that, in any case, follow stomach disorders; an adeno-meningic form, 'whose symptoms indicate an irritation of the mucous membranes of the intestinal duct'; a form occuring above all, in subjects of a lympathic temperament, in women and old men; an adynamic form 'that is manifested above all on the outside by signs of extreme debility and a general atonia of the muscles'. It is probably due to humidity, uncleanliness, visits to hospitals, prisons, and amphitheatres, to bad food, and to the abuse of the venereal pleasures. Lastly, ataxic or malign fever is characterized by 'alternatives of excitation and enfeeblement with the strangest nervous anomalies': it possesses almost the same antecedents as adynamic fever [24].

It is in the very principle of this specification that the paradox resides. In its general form, fever is characterized only by its effects; it has been cut off from any organic substratum; and Pinel does not even mention heat as an essential sign or major symptom of the class of fevers. But when it is a question of dividing up this essence, the function of division is operated by a principle that belongs not to the logical configuration of species, but to the organic spatiality of the body: the blood vessels, the stomach, the intestinal mucous membrane, the muscular or nervous system are called upon in turn to serve as a point of coherence for the formless diversity of the symptoms. And if they can be organized in such a way as to form species, it is not because they are *essential expressions*, but because they are *local signs*. The principle of the essentiality of the fevers has as its concrete, specified content only the possibility of localizing them. From Sauvages's *Nosologie* to Pinel's *Nosographie*, the configuration was reversed: in the first, the local manifestations always carried with them a possible generality; in the second, the general structure envelops the need for a localization.

In these conditions, it is understandable that Pinel should have thought that he could integrate into his symptomatological analysis of fevers the discoveries of Roederer and Wagler, who, in 1783, had shown that mucous fever was always accompanied by traces of internal and external inflammation in the alimentary duct [25]. It is also understandable that he should have accepted the results of Prost's autopsies, which showed obvious intestinal lesions; but it is also understandable why he could not see them himself [26]: for him, the lesional localization occurred of itself, but as a secondary phenomenon, within a symptomatology in which the local signs corresponded not to the seat of the diseases, but to their essence. Finally, it is understandable why Pinel's defenders should have regarded him as the first of the localizers.

He did not limit himself to classifying objects: materializing in some sense a science hitherto overly metaphysical, he tried to localize, if one may be allowed to say so, each disease, or to attribute it with a special seat, that is, to determine the place of its original existence. This idea is evident in the new denominations imposed on the fevers that he continued to call 'essential fevers', as if to pay a final homage to the hitherto dominant ideas, but assigning to each one a particular seat, making the bilious and pituitous fevers, for example, consist of others in the special irritation of certain parts of the intestinal tube [27].

In fact, what Pinel localized was not the diseases, but the signs: and the local value with which they were invested did not indicate a regional origin, an original locus from which the disease derived both birth and form; it simply made it possible to recognize a disease that gave itself this signal as a characteristic symptom of its essence. This being the case, the causal and temporal chain to be established did not proceed from the lesion to the disease, but from the disease to the lesion, as to its consequence and perhaps privileged expression. In 1820, Chomel was still being faithful to the *Nosographie* when he analysed the intestinal ulcerations perceived by Broussais 'as the effect and not the cause of the feverish affection': do they not occur relatively late (on the tenth day of the disease only, when meteorism, sensitivity to the right of the abdomen, and sanious excretions reveal their existence)? Do they not appear in that part of the intestinal duct in which matter that has already been irritated by the disease remains longest (the end of the ileum, caecum, and ascending colon), and in the declivitous segments of the intestine much more frequently than in the vertical, ascending portions [28]? Thus the disease settles in the organism, lays down local signs, and divides itself up throughout the secondary space of the body; but its essential structure remains antecedent. The organic space is provided with references to that structure, it signals it, but does not order it.

The *Examen* of 1816 went to the bottom of Pinel's doctrine in order to denounce its postulates with an astonishing theoretical lucidity. But from the *Histoire des phlegmasies* there was posed as a dilemma what had been thought hitherto to be perfectly compatible: either a fever is idiopathic or it is localizable; and every successful localization will shift the fever away from its status of essentiality.

No doubt this incompatibility, which belonged logically within the anatomo-clinical experience, had been formulated quietly, or at least suspected by Prost when he had shown that the fevers differed from one another according to 'the organ in which the affection occurred', or according to 'the mode of alteration' of the tissues [29], and by Récamier and his pupils when they had studied those diseases that were to be so crucial to the future of medicine, the meningitis group, indicating that 'fevers of this order are rarely essential diseases; they may even always be dependent on an affection of the brain such as phlegmasia, or a serous gathering' [30]. But what enabled Broussais to transform these initial approaches into a systematic form of interpretation for all the fevers was, without the slightest doubt, the diversity and, at the same time, the coherence of the fields of medical experience that he had traversed.

Having been trained just before the Revolution in the medicine of the eighteenth century, having experienced as a medical officer in the navy the problems proper to hospital medicine and to the practice of surgery, having then studied under Pinel and the clinicians of the new École de Santé, and having attended Bichat's lectures and Corvisart's clinical lectures, which introduced him to pathological anatomy, Broussais resumed his military career and followed the army from Utrecht to Mainz, and from Bohemia to Dalmatia, practising like his master Desgenettes comparative medical nosography, and making wide use of the autopsy method. He was familiar with every form of medical experience flourishing towards the end of the eighteenth century; it is not surprising that he was able to derive from these forms as a whole, and from their lines of intersection, the radical lesson that was to give meaning and conclusion to each. Broussais is merely the point of convergence of all these structures, the individually shaped form of their over-all configuration. Indeed, he knew this to be the case, and that in him there spoke

> that observing doctor who will not disdain the experience of others, but who will wish to validate it by his own. . . . Our Schools of Medicine, which have succeeded in freeing themselves from the yoke of the old system, and in preserving themselves from the contagion of the new, have, for some years now, been

training subjects capable of giving confidence to the still-tottering step of the curing art. Living among their fellow-citizens or scattered afar in our armies, they observe and meditate. . . . One day, perhaps, they will make their voices heard [31].

On his return to Dalmatia in 1808, Broussais published his *Histoire des phlegmasies chroniques.*

This represents a sudden return to the pre-clinical idea that fever and inflammation belong to the same pathological process. But whereas in the eighteenth century this identity rendered the distinction between general and local a secondary one, in Broussais it is the natural consequence of Bichat's tissual principle, that is, of the need to find the surface of organic attack. Each tissue has its own mode of alteration: it is, therefore, by analysis of the particular forms of inflammation at the level of the areas of the organism that one must begin the study of what are known as the fevers. There are inflammations in those tissues that possess a great many capillary blood vessels (such as the pia mater or the pulmonary lobes), which cause a strong thermal thrust, an alteration of the nervous functions, a disturbance to the secretions, and possible muscular disorders (agitation, contractions); those tissues possessing few red capillaries (thin membranes) undergo similar, but slighter disorders; lastly, the inflammation of the lymphatic vessels causes disturbances in nutrition and in the serous secretions [32].

Against the background of this quite encompassing specification, which is very close in style to Bichat's analyses, the world of fevers becomes strangely simplified. One will now find in the lung only those phlegmasias that correspond to the first type of inflammation (catarrh and pleuropneumonia), those deriving from the second type (pleurisy), and those that originate in an inflammation of the lymphatic vessels (tubercular phthisis). In the case of the digestive system, the mucous membrane may be affected either at the level of the stomach (gastritis) or in the intestine (enteritis, peritonitis). They evolve in a convergent manner, according to the logic of tissual propagation: when an inflammation of the blood persists, it always reaches the lymphatic vessels; that is why phlegmasias of the respiratory system 'all culminate in pulmonary phthisis' [33]; while intestinal inflammations usually tend to ulcerations of the peritoneum. Homogeneous in origin, and convergent in their terminal forms, the phlegmasias proliferate in multiple symptoms only in the interval between the two. By way of sympathy they reach new regions and new tissues: they may either take the form of a progression along the stages of organic life (thus, inflammation of the intestinal mucous membrane may alter the bilious and urinary secretions or cause spots on the skin or coated mouth) or attack in turn the relating functions (headaches, muscular pains, dizzy spells, torpor, delirium). Thus all the symptomatological varieties may originate in this generalization.

It is here that the great conceptual conversion that Bichat's method had authorized, but not yet clarified, resides: in becoming generalized, the local disease produces the symptoms particular to each species; but in its first, geographical form, fever is merely a locally individualized phenomenon with a general pathological structure. In other words, the particular symptom (nervous or hepatic) is not a local sign; on the contrary, it is an index of generalization; only the general symptom of inflammation bears within itself the need for a localized attack-point. Bichat's preoccupation remained that of finding an organic base for general diseases: hence his search for organic universalities. Broussais dissociates doublets, a particular symptom—a local lesion, a general symptom—and an over-all alteration, intersects their elements, and shows the over-all alteration in the particular symptom, the geographical lesion in the general symptom. From now on, the organic space of the localization is really independent of the space of the nosological configuration: the second slips over the first, shifts its values in relation to it, and reflects them only at the price of an inverted projection.

But what is inflammation, a process of general structure, but with an always localized attack-point? The old symptomatic analysis characterized it by tumour, soreness, heat, pain—which does not correspond to the forms that it assumes in the tissues; the inflammation of a membrane presents neither pain nor heat, still less soreness. Inflammation is not a constellation of signs: it is a process that develops within a tissue: 'any local agitation of the organic movements large enough to disturb the harmony of the functions, and to disorganize the tissue in which it is fixed, must be regarded as inflammation' [34]. It is, therefore, a phenomenon involving two pathological layers at different levels and with different chronologies: first an attack on the functions, then an attack on the texture. Inflammation has a physiological reality that may anticipate anatomical disorganization, which makes it perceptible to the eyes. Hence the need for a physiological medicine, 'observing life, not abstract life, but the life of the organs, and in the organs, in relation with all the agents that may exert influence over them' [35]; pathological anatomy, conceived as a mere examination of lifeless bodies, is its own limit, while ever 'the role and sympathies of all the organs are far from being perfectly known' [36].

In order to detect this primary, fundamental, functional disorder, the gaze must be able to detach itself from the lesional site, for it is not given at the outset, although the disease, in its original source, was always localizable; indeed, it has to locate that organic root before the lesion, by means of the functional disorders and their symptoms. It is here that symptomatology rediscovers its role, but it is a role based entirely on the local character of the pathological attack: by returning along the path of organic sympathies and influences, it must, beneath the endlessly extended network of symptoms, 'induce' or 'deduce' (Broussais uses both words in the same sense) the initial point of physiological disturbance. 'To study the altered organs without referring to the symptoms of the diseases is like regarding the stomach independently of the digestion' [37]. Thus, instead of praising the advantages of description, as is all too commonly the case, while depreciating 'induction as no more than hypothetical theory, a priori systematizing of vain conjectures' [38], one will make the observation of symptoms speak the very language of pathological anatomy.

This represents a new organization of the medical gaze in relation to Bichat: since the *Traité des membranes*, the principle of visibility had been an absolute rule, and localization was merely its consequence. With Broussais, the order is inverted; it is because disease, in its nature, is local that it is, in a secondary way, visible. Broussais, above all in the *Histoire des phlegmasies*, admits (and in doing so he goes further than Bichat, for whom the vital diseases need not necessarily leave a trace) that every 'pathological affection' implies 'a particular modification to the phenomenon that restores our bodies to the laws of inorganic matter': as a result, 'if corpses have sometimes seemed to us to be silent, it is because we were ignorant of the art of questioning them' [39]. But when the attack is of an especially physiological kind, these alterations may be scarcely perceptible; or, again, they may, like the spots on the skin in intestinal disorders, disappear with death; they may be, at least in extension and perceptual importance, out of all proportion to the disorder that they cause: the important thing, in fact, is not what can be *seen* of these alterations, but what, in them, is determined by the *place* in which they develop. By knocking down the noso-

logical wall maintained by Bichat between the vital or functional
disorder and the organic alteration, Broussais, because of an obvious
structural necessity, gave precedence to the axiom of localization
over the principle of visibility. Disease exists *in space* before it
exists *for sight*. The disappearance of the two great a priori classes
of nosology opened up for medicine an entirely spatial field of
investigation, determined throughout by these local values. It is
curious to observe that this absolute spatialization of medical experience is due not to the definitive integration of normal and pathological *anatomy*, but to the first effort to define a *physiology* of the
morbid phenomenon.

But we must go back further still into the constituent elements
of this new medicine, and pose the question of the origin of inflammation. Inflammation being a local excitation of organic movements, it presupposes in the tissues a certain 'aptitude to be moved'
and, in contact with these tissues, an agent that arouses and exag-
gerates the mechanisms. Such an agent is irritability, 'a faculty that
tissues possess of moving when brought into contact with a foreign
body. . . . Haller attributed this property only to the muscles;
but it is now agreed that it is common to all the tissues' [40]. It must
not be confused with sensitivity, which is 'an awareness of the
movements caused by foreign bodies', and forms only an additional,
secondary phenomenon in relation to irritability: the embryo is not
yet sensitive, the apoplectic no longer is; both are irritable. Increase
in irritative action is caused 'by bodies or objects, living or not
living' [41], coming into contact with tissues; they may therefore
be internal or external agents, but they are in any case foreign to
the functioning of the organ; the serosity of one tissue may become
irritating for another or for itself if it is too abundant, or if there is
a change of climate or regimen. An organism is sick only in relation
to the solicitations of the external world, or of alterations in its
functioning or anatomy. 'After many hesitant steps, medicine is
pursuing at last the only road that can lead it to the truth: observation of the relations between man and external modifications,
and between men's organs' [42].

By means of this conception of the external agent or of internal
modification, Broussais avoided one of the themes that had dominated medicine, with few exceptions, since Sydenham: the impos-
sibility of defining the cause of diseases. From this point of view,
nosology from Sauvages to Pinel had been like a figure confined

within this abandonment to causal assignation: the disease set in and
flourished in its essential affirmation, and causal series were merely
so many elements within a schema in which the nature of the
pathological served it as an effective cause. With Broussais—which
was not yet the case with Bichat—localization demands an enveloping causal schema: the seat of the disease is merely the link point of
the irritating cause, a point that is determined by both the irritability
of the tissue and the irritating power of the agent. The local space
of the disease is also, immediately, a causal space.

And so—and this was the great discovery of 1816—the *being*
of the disease disappears. As an organic reaction to an irritating
agent, the pathological phenomenon can no longer belong to a
world in which the disease, in its particular structure, would exist
in conformity with a dominant type that preceded it, and in which
it was fulfilled, once individual variations and non-essential accidents had been set aside; it is caught up in an organic web in which
the structures are spatial, the determinations causal, the phenomena
anatomical and physiological. Disease is now no more than a certain
complex movement of tissues in reaction to an irritating cause: it
is in this that the whole essence of the pathological lies, for there
are no longer either essential diseases or essences of diseases. 'All
classifications that tend to make us regard diseases as particular
beings are defective, and a judicious mind is constantly, almost in
spite of itself, drawn towards a search for sick organs' [43]. Thus
fever cannot be essential: it is 'no more than an acceleration in the
flow of blood . . . accompanied by an increase of calorification and
a lesion of the principal functions. This state of the economy is
always dependent on a local irritation' [44]. All the fevers are
dissolved into one long organic process, a theory was proposed
almost in its entirety in the text of 1808 [45], affirmed in 1816, and
outlined once more eight years later in the *Catéchisme de la
Médecine physiologique*. At the origin of all the fevers lay a single
gastro-intestinal irritation: first, a simple redness, then wine-coloured
spots that become more and more numerous in the ileo-caecal
region; these spots often develop into swollen areas, ultimately
leading to ulcerations. On this constant anatomo-pathological web,
which defines the origin and general form of gastro-enteritis, the
processes proliferate: when irritation of the digestive canal has
spread more in extent than in depth, it causes a considerable bile
secretion, and pain in the locomotor muscles: this is what Pinel

called bilious fever; in a lymphatic subject, or when the intestine is filled with mucous, gastro-enteritis takes on the form that was known as mucous fever; what was called adynamic fever 'is simply gastro-enteritis that has reached such a degree of intensity that there is a reduction in strength, and a blunting of the intellectual powers . . . the tongue becomes brown, and the mouth is coated with a blackish substance'; when the irritation reaches by sympathy to the investing membrane of the brain, the fevers take on 'malign' forms [46]. Through these and other branches, gastro-enteritis gradually spreads throughout the whole organism: 'It is certainly true that the flow of blood is precipitated into all the tissues; but this does not prove that the cause of these phenomena resides in all the points of the body' [47]. So fever has to be deprived of its status as a general state, and 'disessentialized' [48], in favour of physio-pathological processes that specify its manifestations.

This dissolution of the ontology of fever, together with the errors that it involved (at a period when the difference between meningitis and typhus was beginning to be seen clearly), is the best-known element of the analysis. In fact, in the general structure of its analysis, it is merely the negative counterpart of a positive, and much more subtle, element: the idea of a medical (anatomical and, above all, physiological) method applied to organic illness: one must 'seek in physiology the characteristic features of diseases, and by skilful analysis disentangle the often confused cries of the sick organs' [49]. This medicine of the sick organs involves three stages:

1. Decide which organ is sick, which can be done on the basis of the symptoms manifested, so long as one knows 'all the organs, all the tissues that make up the means of communication by which these organs are linked together, and the changes that a modification in one organ may bring about in others';

2. 'Explain how an organ became sick', by means of an external agent; by taking account of the essential fact that irritation may cause hyperactivity or, on the contrary, functional asthenia, and that 'these two modifications almost always exist together in our economy' (under the effect of cold, the activity of the cutaneous secretions diminishes, while that of the lung increases);

3. 'Indicate what is to be done for the illness to cease': that is, eliminate not only the cause (cold in pneumonia), but also 'the effects that do not always disappear when the cause has ceased to operate' (congestion of the blood maintains irritation in the lungs of pneumonics) [50].

In the critique of medical 'ontology', the notion of organic 'sickness' goes further and more deeply perhaps than that of irritation. Irritation still involved an abstract structure: the universality that enabled it to explain everything formed for the gaze directed upon the organism a final screen of abstraction. The notion of a 'sickness' of the organs involved only the idea of a relationship of the organ with an agent or an environment, that of a reaction to attack, that of an abnormal functioning, and, finally, that of the disturbing influence of the element attacked upon the other organs. Henceforth the medical gaze will be directed only upon a space filled with the forms of composition of the organs. The space of the disease is, without remainder or shift, the very space of the organism.

The medicine of diseases has come to an end; there now begins a medicine of pathological reactions, a structure of experience that dominated the nineteenth century, and, to a certain extent, the twentieth, since the medicine of pathogenic agents was to be contained within it, though not without certain methodological modifications.

So necessary was Broussais's attempt in the development and balancing of structures that it caused a shift in the whole of medical experience. We may leave aside the endless discussions that set Broussais's disciples against the last followers of Pinel. The anatomopathological analyses carried out by Petit and Serres on enteromesenteric fever [51], the distinction re-established by Caffin between thermic symptoms and the so-called feverish diseases [52], the work of Lallemand on acute cerebral affections [53], and, lastly, Bouillaud's *Traité*, devoted to 'the so-called essential fevers' [54], gradually rendered unproblematic the very thing that continued to feed controversy. In the end, the controversy died down. Chomel, who in 1821 affirmed the existence of general fevers without lesions, recognized in 1834 that they all had an organic localization [55]; Andral had devoted a volume in the first edition of his *Clinique médicale* to the classification of fevers; in the second edition, he divided them into phlegmasias of the viscera and phlegmasias of the nervous centres [56].

And yet, to the end of his life, Broussais was the object of passionate attack; and after his death, his reputation continued to decline. It could hardly be otherwise. Broussais succeeded in circumventing the idea of essential diseases only at an extraordinarily high price; he had had to re-arm the old, much criticized (and justly criticized by pathological anatomy) notion of sympathy; he had had to return to the Hallerian concept of irritation; he had fallen back on a pathological monism reminiscent of Brown, and brought back into play, in the logic of his system, the old practice of bleeding. All these reversions had been structurally necessary if a medicine of organs was to appear in all its purity and if medical perception was to be liberated from all nosological prejudice. But by virtue of that very fact it incurred the risk of losing itself in both the diversity of phenomena and the homogeneity of the process. Before fixing the inevitable ordering on which all singularities were based, perception swung between monotonous irritation and the endless violence 'of the cries of sick organs': lancet and leech.

Everything was justified in the frenzied attacks that Broussais's contemporaries launched against him. And yet not entirely so: it was to his 'physiological medicine' that they owed this anatomo-clinical perception, conquered at last in its totality and capable of self-correction, this perception in the name of which they were right and he wrong, or at least its definitive form of balance. Everything in Broussais ran counter to his time, but he had fixed for his period the final element of *the way to see*. Since 1816, the doctor's eye has been able to confront a sick organism. The historical and concrete a priori of the modern medical gaze was finally constituted.

The decipherment of structures merely brought about a series of rehabilitations. But since there are still doctors, and others, who think they are practising history when they write biographies and hand out praise and blame, here, for them, is a text written by a doctor who was not so ignorant: 'The publication of the *Examen de la doctrine médicale* is one of those important events that will be long remembered in the annals of medicine. . . . The medical revolution of which M. Broussais laid the foundations in 1816 is undoubtedly the most remarkable that medicine has undergone in modern times' [57].

NOTES

[1] M.-A. Petit, *Traité de la fièvre entéro-mésentérique* (Paris, 1812, pp. 147-8).

[2] A.-F. Chomel, *Éléments de pathologie générale* (Paris, 1817, p. 523).

[3] Bichat, *Anatomie générale*, vol. I, p. xcviii.

[4] Laënnec, article on 'Anatomie pathologique', *Dictionnaire des Sciences médicales*, vol. II, p. 47.

[5] Bayle, 2nd article on 'Anatomie pathologique', *ibid.*, p. 62.

[6] J. Cruveilhier, *Essai sur l'anatomie pathologique* (Paris, 1816, I, pp. 21–4).

[7] Cf. above, Chapter 1, p. 14.

[8] Cf. above, Chapter 7, p. 117.

[9] P. A. Prost relates how he showed Corvisart and Pinel 'inflammations and alterations of the internal membrane of the intestines the existence of which they so little suspected that they had allowed corpses in which they were pointed out to them to pass through their hands without opening up the intestines' (*Traité de cholera-morbus*, 1832, p. 30).

[10] X. Bichat, *Anatomie descriptive*, vol. I, p. 19.

[11] Laënnec, *Traité de l'auscultation*, preface, p. xx.

[12] In a dissertation that was awarded a prize, Martin criticizes the oversimple use that is made of the term 'disease', which he would like to confine to affections resulting from a nutritional defect of the tissues. Cf. *Bulletin des sciences médicales*, vol. V (1810), pp. 167–88.

[13] H. Mondor, *Vie de Dupuytren* (Paris, 1945, p. 176): 'a doctor drunk with his own histrionics . . . a vain, loud-mouthed charlatan . . . his tricks, his impudence, his verbose argumentativity, his high-sounding errors . . . his illusionist's self-possession.'

[14] Boerhaave, *Aphorisme*.

[15] Stahl, quoted in Dagoumer, *Précis historique de la fièvre* (Paris, 1831, p. 9).

[16] Quoted *ibid.*

[17] Apart from a few variations, this schema is to be found in Boerhaave (*Aphorismes*, 563, 570, 581), Hoffmann (*Fundamenta Medica*), Stoll (*Aphorismes sur la connaissance et la curation des fièvres*), Huxham (*Essai sur les fièvres*), and Boissier de Sauvages (*Nosologie méthodique*, vol. II).

[18] Huxham, *Essai sur les fièvres* (Fr. trans., Paris, 1752, p. 339).

[19] Stoll, 'Aphorisme sur la connaissance et la curation des fièvres', *Encyclopédie des Sciences médicales*, 7th division, vol. V, p. 347.

[20] Grimaud, *Traité des fièvres* (Montpellier, 1791, vol. I, p. 89).

[21] Bouillaud provides a very clear analysis of this in *Traité des fièvres dites essentielles* (Paris, 1826, p. 8).

[22] Morgagni, *De sedibus et causis morborum*, Epist. 49, art. 5.

[23] Pinel, *Nosographie philosophique*, 5th edn., 1813, I, p. 320.

[24] *Ibid.*, pp. 9–10 and pp. 323–4.

[25] Roederer and Wagler, *De morbo mucoso* (Göttingen, 1783).

[26] Cf. above, n. 9.

[27] Richerand, *Histoire de la chirurgie* (Paris, 1825, p. 250).

[28] A.-F. Chomel, *De l'existence des fièvres essentielles* (Paris, 1820, pp. 10–12).

[29] Prost, *La médecine des corps éclairée par l'ouverture et l'observation* (Paris, Year XII, vol. I, pp. xxii and xxiii).

[30] P.-A. Dan de La Vautrie, *Dissertation sur l'apoplexie considérée spécialement comme l'effet d'une phlegmasie de la substance cérébrale* (Paris, 1807).

[31] F.-J.-V. Broussais, *Histoire des phlegmasies chroniques*, vol. II, pp. 3-4.

[32] *Ibid.*, vol. I, pp. 55-6.

[33] *Ibid.*, vol. I, preface, p. xiv.

[34] *Ibid.*, vol. I, p. 6.

[35] Broussais, *Sur l'influence que les travaux des médecins physiologistes ont exercée sur l'état de la médecine* (Paris, 1832, pp. 19-20).

[36] Broussais, *Examen des doctrines* (2nd edn., Paris, 1821, vol. II, p. 647).

[37] *Ibid.*, p. 671.

[38] Broussais, *Mémoire sur la philosophie de la médecine* (Paris, 1832, pp. 14-15).

[39] Broussais, *Histoire des phlegmasies*, vol. I, preface, p. v.

[40] Broussais, *De l'irritation et de la folie* (Paris, 1839 edn., vol. I, p. 3).

[41] *Ibid.*, p. 1, n. 1.

[42] *Ibid.*, preface to 1828 edition, 1839 end., vol. I, p. lxv.

[43] Broussais, *Examen de la doctrine* (Paris, 1816; preface).

[44] *Ibid.*, 1821 edn., p. 399.

[45] In 1808, Broussais still set aside the malign typhus diseases (ataxic fevers), for which he had not discovered visceral inflammation at autopsy (*Examen des doctrines*, 1821, vol. II, pp. 666-8).

[46] Broussais, *Catéchisme de la médecine physiologiste* (Paris, 1824, pp. 28-30).

[47] *Examen des doctrines*, 1821, vol. II, p. 399.

[48] The expression is to be found in Broussais's replay to Fodera ('Histoire de quelques doctrines médicales'), *Journal universel des Sciences médicales*, vol. XXIV.

[49] Broussais, *Examen de la doctrine*, 1816, preface.

[50] *Examen des doctrines*, 1821, vol. I, pp. 52-5. In the text on *L'influence des médecins physiologistes*, 1832, Broussais adds, between the 2nd and 3rd precepts, that of determining the action of the sick organ upon the other organs.

[51] M.-A. Petit, and Serres, *Traité de la fièvre entéro-mésentérique* (Paris, 1813).

[52] Caffin, *Traité analytique des fièvres essentielles* (Paris, 1811).

[53] Lallemand, *Recherches anatomo-pathologiques sur l'encéphale* (Paris, 1820).

[54] Bouillaud, *Traité clinique ou expérimental des fièvres dites essentielles* (Paris, 1826).

[55] Chomel, *Traité des fièvres et des maladies pestilentielles*, 1821, and *Leçons sur la fièvre typhoïde*, 1834.

[56] Andral, *Clinique médicale* (Paris, 1823-1827, 4 vols.). According to one story, Pinel had wanted to cut the classification of fevers from the last edition of the *Nosologie*, but was prevented from doing so by his publisher.

[57] Bouillaud, *Traité des fièvres dites essentielles* (Paris, 1826, p. 13).

Conclusion

This book is, among others, an attempt to apply a method in the confused, under-structured, and ill-structured domain of the history of ideas.

Its historical support is limited since it deals, on the whole, with the development and methods of medical observation over less than half a century. Yet it concerns one of those periods that mark an ineradicable chronological threshold: the period in which illness, counter-nature, death, in short, the whole dark underside of disease came to light, at the same time illuminating and eliminating itself like night, in the deep, visible, solid, enclosed, but accessible space of the human body. What was fundamentally invisible is suddenly offered to the brightness of the gaze, in a movement of appearance so simple, so immediate that it seems to be the natural consequence of a more highly developed experience. It is as if for the first time for thousands of years, doctors, free at last of theories and chimeras, agreed to approach the object of their experience with the purity of an unprejudiced gaze. But the analysis must be turned around: it is the forms of visibility that have changed; the new medical spirit to which Bichat is no doubt the first to bear witness in an absolutely coherent way cannot be ascribed to an act of psychological and epistemological purification; it is nothing more than a syntactical reorganization of disease in which the limits of the visible and invisible follow a new pattern; the abyss beneath illness, which was the illness itself, has emerged into the light of language —the same light, no doubt, that illuminates the *120 Journées de Sodome, Juliette,* and the *Désastres de Soya* [1].

But we are concerned here not simply with medicine and the way in which, in a few years, the particular knowledge of the individual patient was structured. For clinical experience to become possible as a form of knowledge, a reorganization of the hospital field, a new definition of the status of the patient in society, and the establishment of a certain relationship between public assistance and medical experience, between help and knowledge, became necessary; the patient has to be enveloped in a collective, homogeneous space. It was also necessary to open up language to a whole new domain: that of a perpetual and objectively based correlation of the visible and the expressible. An absolutely new use of scientific discourse was then defined: a use involving fidelity and unconditional subservience to the coloured content of experience—to say what one sees; but also a use involving the foundation and constitution of experience—showing by saying what one sees. It was necessary, then, to place medical language at this apparently superficial but in fact very deeply embedded level at which the descriptive formula is also a revealing gesture. And this revelation in turn involved as its field of origin and of manifestation of truth the discursive space of the corpse: the interior revealed. The constitution of pathological anatomy at the period when the clinicians were defining their method is no mere coincidence: the balance of experience required that the gaze directed upon the individual and the language of description should rest upon the stable, visible, legible basis of death.

This structure, in which space, language, and death are articulated—what is known, in fact, as the anatomo-clinical method—constitutes the historical condition of a medicine that is given and accepted as positive. Positive here should be taken in the strong sense. Disease breaks away from the metaphysic of evil, to which it had been related for centuries; and it finds in the visibility of death the full form in which its content appears in positive terms. Conceived in relation to nature, disease was the non-assignable negative of which the causes, forms, and manifestations were offered only indirectly and against an ever-receding background; seen in relation to death, disease becomes exhaustively legible, open without remainder to the sovereign dissection of language and of the gaze. It is when death became the concrete a priori of medical experience that death could detach itself from counter-nature and become *embodied* in the *living bodies* of individuals.

CONCLUSION 197

It will no doubt remain a decisive fact about our culture that its first scientific discourse concerning the individual had to pass through this stage of death. Western man could constitute himself in his own eyes as an object of science, he grasped himself within his language, and gave himself, in himself and by himself, a discursive existence, only in the opening created by his own elimination: from the experience of Unreason was born psychology, the very possibility of psychology; from the integration of death into medical thought is born a medicine that is given as a science of the individual. And, generally speaking, the experience of individuality in modern culture is bound up with that of death: from Hölderlin's Empedocles to Nietzsche's Zarathustra, and on to Freudian man, an obstinate relation to death prescribes to the universal its singular face, and lends to each individual the power of being heard forever; the individual owes to death a meaning that does not cease with him. The division that it traces and the finitude whose mark it imposes link, paradoxically, the universality of language and the precarious, irreplaceable form of the individual. The sense-perceptible, which cannot be exhausted by description, and which so many centuries have wished to dissipate, finds at last in death the law of its discourse; it is death that fixes the stone that we can touch, the return of time, the fine, innocent earth beneath the grass of words. In a space articulated by language, it reveals the profusion of bodies and their simple order.

It is understandable, then, that medicine should have had such importance in the constitution of the sciences of man—an importance that is not only methodological, but ontological, in that it concerns man's being as object of positive knowledge.

The possibility- for the individual of being both subject and object of his own knowledge implies an inversion in the structure of finitude. For classical thought, finitude had no other content than the negation of the infinite, while the thought that was formed at the end of the eighteenth century gave it the powers of the positive: the anthropological structure that then appeared played both the critical role of limit and the founding role of origin. It was this reversal that served as the philosophical condition for the organization of a positive medicine; inversely, this positive medicine marked, at the empirical level, the beginning of that fundamental relation that binds modern man to his original finitude. Hence the funda-

mental place of medicine in the over-all architecture of the human
sciences: it is closer than any of them to the anthropological structure that sustains them all. Hence, too, its prestige in the concrete
forms of existence: health replaces salvation, said Guardia. This
is because medicine offers modern man the obstinate, yet reassuring
face of his finitude; in it, death is endlessly repeated, but it is also
exorcized; and although it ceaselessly reminds man of the limit that
he bears within him, it also speaks to him of that technical world
that is the armed, positive, full form of his finitude. At that point
in time, medical gestures, words, gazes took on a philosophical
density that had formerly belonged only to mathematical thought.
The importance of Bichat, Jackson, and Freud in European culture
does not prove that they were philosophers as well as doctors, but
that, in this culture, medical thought is fully engaged in the philosophical status of man.

This medical experience is therefore akin even to a lyrical
experience that his language sought, from Hölderlin to Rilke. This
experience, which began in the eighteenth century, and from which
we have not yet escaped, is bound up with a return to the forms
of finitude, of which death is no doubt the most menacing, but also
the fullest. Hölderlin's Empedocles, reaching, by voluntary steps,
the very edge of Etna, is the death of the last mediator between
mortals and Olympus, the end of the infinite on earth, the flame
returning to its native fire, leaving as its sole remaining trace that
which had precisely to be abolished by his death: the beautiful,
enclosed form of individuality; after Empedocles, the world is
placed under the sign of finitude, in that irreconcilable, intermediate
state in which reigns the Law, the harsh law of limit; the destiny
of individuality will be to appear always in the objectivity that
manifests and conceals it, that denies it and yet forms its basis: 'here,
too, the subjective and the objective exchange faces'. In what at
first sight might seem a very strange way, the movement that sustained lyricism in the nineteenth century was one and the same as
that by which man obtained positive knowledge of himself; but is
it surprising that the figures of knowledge and those of language
should obey the same profound law, and that the irruption of
finitude should dominate, in the same way, this relation of man to
death, which, in the first case, authorizes a scientific discourse in a
rational form, and, in the second, opens up the source of a language
that unfolds endlessly in the void left by the absence of the gods?

The formation of clinical medicine is merely one of the more visible witnesses to these changes in the fundamental structures of experience; it is obvious that these changes go well beyond what might be made out from a cursory reading of positivism. But when one carries out a vertical investigation of this positivism, one sees the emergence of a whole series of figures—hidden by it, but also indispensable to its birth—that will be released later, and, paradoxically, used against it. In particular, that with which phenomenology was to oppose it so tenaciously was already present in its underlying structures: the original powers of the perceived and its correlation with language in the original forms of experience, the organization of objectivity on the basis of sign values, the secretly linguistic structure of the datum, the constitutive character of corporal spatiality, the importance of finitude in the relation of man with truth, and in the foundation of this relation, all this was involved in the genesis of positivism. Involved, but forgotten to its advantage. So much so that contemporary thought, believing that it has escaped it since the end of the nineteenth century, has merely rediscovered, little by little, that which made it possible. In the last years of the eighteenth century, European culture outlined a structure that has not yet been unraveled; we are only just beginning to disentangle a few of the threads, which are still so unknown to us that we immediately assume them to be either marvellously new or absolutely archaic, whereas for two hundred years (not less, yet not much more) they have constituted the dark, but firm web of our experience.

NOTE

[1] All works by the Marquis de Sade.

Bibliography

I. NOSOLOGY

ALIBERT, J.-L., *Nosologie naturelle* (Paris, 1817).
BOISSIER DE SAUVAGES, Fr., *Nosologie méthodique* (Fr. trans., Lyons, 1772, 10 vols.).
CAPURON, J., *Nova medicinae elementa* (Paris, 1804).
Ch . . . , J.-J., *Nosographiae compendium* (Paris, 1816).
CHAUSSIER, Fr., *Table générale des méthodes nosologiques* (Paris, n.d.).
CULLEN, W., *Apparatus ad nosologiam methodicam* (Amsterdam, 1775).
———, *Institutions de médecine pratique* (Fr. trans., Paris, 1785, 2 vols.).
DUPONT, J.-Ch., *Y a-t-il de la différence dans les systèmes de classification dont on se sert avec avantage dans l'étude de l'histoire naturelle et ceux qui peuvent être profitables à la connaissance des maladies?* (Bordeaux, 1803).
DURET, F.-J.-J., *Tableau d'une classification générale des maladies* (Paris, 1813).
FERCOQ, G.A., *Synonymie ou concordance de la nomenclature de la Nosographie philosophique du Pr Pinel avec les anciennes nosologies* (Paris, 1812).
FRANK, J. P., *Synopsis nosologiae methodicae* (Ticini, 1790).
LATOUR, F.-D., *Nosographie synoptique* (Paris, 1810, 1st vol. only).
LINNAEUS, C., *Genera morborum* (Fr. trans., SAUVAGES, cf. above).
PINEL, Ph., *Nosographie philosophique* (Paris, year VI).
SAGAR, J. B. M., *Systema morborum systematicum* (Vienna, 1771).
SYDENHAM, Th., *Médecine pratique* (Fr. trans., Paris, 1784).
VOULONNE, *Déterminer les maladies dans lesquelles la médecine agissante est préférable à l'expectante* (Avignon, 1776).

II. MEDICAL ADMINISTRATION AND GEOGRAPHY

AUDIN-ROUVIÉRE, J.-M., *Essai sur la topographie physique et médicale de Paris* (Paris, Year II).
BACHER, A., *De la médecine considérée politiquement* (Paris, Year IX).

BANAU and TURBEN, *Mémoires sur les épidémies du Languedoc* (Paris, 1766).

BARBERET, D., *Mémoire sur les maladies épidémiques des bestiaux* (Paris, 1766).

BIENVILLE, J.-D.-T., *Traité des erreurs populaires sur la médecine* (The Hague, 1775).

CATTET, J.-J. and GARDET, J.-B., *Essai sur la contagion* (Paris, Year II).

CERVEAU, M., *Dissertation sur la médecine des casernes* (Paris, 1803).

CLERC, *De la contagion* (St. Petersburg, 1771).

COLOMBIER, J., *Préceptes sur la santé des gens de guerre* (Paris, 1775).

———, *Code de médecine militaire* (5 vols., Paris, 1772).

DAIGNAN, G., *Ordre du service des hôpitaux militaires* (Paris, 1785).

———, *Tableau des variétés de la vie humaine* (2 vols., Paris, 1786).

———, *Centuries médicales du XIXᵉ siècle* (Paris, 1807–1808).

———, *Conservatoire de Santé* (Paris, 1802).

DESGENETTES, R.-N., *Histoire médicale de l'armée d'Orient* (Paris, 1802).

———, *Opuscules* (Cairo, n.d.).

FOUQUET, H., *Observations sur la constitution des six premiers mois de l'an V à Montpellier* (Montpellier, Year VI).

FRANK, J.-P., *System einer vollständigen medizinischen Polizei* (4 vols., Mannheim, 1779–1790).

FRIER, F., *Guide pour la conservation de l'homme* (Grenoble, 1789).

GACHET, L.-E., *Problème médico-politique pour ou contre les arcanes* (Paris, 1791).

GACHET, M., *Tableau historique des événements présents relatif à leur influence sur la santé* (Paris, 1790).

GANNE, A., *L'homme physique et moral* (Strasbourg, 1791).

GUINDANT, T., *La nature opprimée par la médecine moderne* (Paris, 1768).

GUYTON-MORVEAU, L.-B., *Traité des moyens de désinfecter l'air* (Paris, 1801).

HAUTESIERCK, F.-M., *Recueil d'observations de médecine des hôpitaux militaires* (2 vols., Paris, 1766–1772).

HILDENBRAND, J.-V., *Du typhus contagieux* (Fr. trans., Paris, 1811).

HORNE, D. R. DE, *Mémoire sur quelques objets qui intéressent plus particulièrement la salubrité de la ville de Paris* (Paris, 1788).

Instruction sur les moyens d'entretenir la salubrité et de purifier l'air des salles dans le hôpitaux militaires (Paris, Year II).

JACQUIN, A.-P., *De la Santé* (Paris, 1762).

LAFON, J.-B., *Philosophie médicale* (Paris, 1796).

LANTHENAS, F., *De l'influence de la liberté sur la santé, la morale et le bonheur* (Paris, 1798).

LAUGIER, E.-M., *L'art de faire cesser la peste* (Paris, 1784).

LEBÈGUE DE PRESLE, *Le conservateur de Santé* (Paris, 1772).

LEBRUN, *Traité théorique sur les maladies épidémiques* (Paris, 1776).

LEPECQ DE LA CLOTURE, L., *Collection d'observations sur les maladies et constitutions épidémiques* (2 vols., Rouen, 1778).

LIOULT, P.-J., *Les charlatans dévoilés* (Paris, Year VIII).

MACKENZIE, J., *Histoire de la santé et de l'art de la conserver* (The Hague, 1759).

MARET, M., *Quelle influence les mœurs des Français ont sur leur santé* (Amiens, 1772).

Médecine militaire ou Traité des maladies tant internes qu'externes auxquelles les militaires sont exposés pendant la paix ou la guerre (6 vols., Paris, 1778).

MENURET, J.-J., *Essai sur l'action de l'air dans les maladies contagieuses* (Paris, 1781).

———, *Essai sur l'histoire médico-topographique de Paris* (Paris, 1786).

MURAT, J.-A., *Topographie médicale de la ville de Montpellier* (Montpellier, 1810).

NICOLAS, P.-F., *Mémoires sur les maladies épidémiques qui ont régné dans la province de Dauphiné* (Grenoble, 1786).

PETIT, M.-A., *Sur l'influence de la Révolution sur la santé publique* (1796).

———, in *Essai sur la médecine du cœur* (Lyon, 1806).

PICHLER, J.-F.-C., *Mémoire sur les maladies contagieuses* (Strasbourg, 1786).

Préceptes de santé ou Introduction au Dictionnaire de Santé (Paris, 1772).

QUATROUX, Fr., *Traité de la peste* (Paris, 1771).

RAZOUX, J., *Tables nosologiques et météorologiques dressées à l'Hôtel-Dieu de Nîmes* (Basel, 1767).

Réflexions sur le traitement et la nature des épidémies lues à la Société royale de Médecine le 27 mai 1785 (Paris, 1785).

ROY-DESJONCADES, A., *Les lois de la nature applicables aux lois physiques de la médecine* (2 vols., Paris, 1788).

ROCHARD, C.-C.-T., *Programme de cours sur les maladies épidémiques* (Strasbourg, Year XIII).

RUETTE, F., *Observations cliniques sur une maladie épidémique* (Paris, n.d.).

SALVERTE, E., *Des rapports de la médecine avec la politique* (Paris, 1806).

SOUQUET, *Essai sur l'histoire topographique médico-physique du district de Boulogne* (Boulogne, Year II).

TALLAVIGNES, J.-A., *Dissertation sur la médecine où l'on prouve que l'homme civilisé est plus sujet aux maladies graves* (Carcassonne, 1821).

THIERY, *Vœux d'un patriote sur la médecine en France* (Paris, 1789).

III. REFORM OF PRACTICE AND TEACHING

BARAILLON, J.-F., *Rapport sur la partie de police qui tient à la médecine, 8 germ. an VI* (Paris, Year VI).

———, *Opinion sur le projet de la commission d'Instruction publique relatif aux Ecoles de Médecine, 7 germ. an VI* (Paris, Year VI).

BAUMES, J.-B.-J., *Discours sur la nécessité des sciences dans une nation libre* (Montpellier, Year III).

CABANIS, P.-J.-G., *Œuvres* (2 vols., Paris, 1956).

CALÈS, J.-M., *Projet sur les Ecoles de santé, 12 prairial an V* (Paris, Year V).

———, *Opinion sur les Ecoles de Médecine, 17 germinal an VI* (Paris, Year VI).

CANTIN, D.-M.-J., *Projet de réforme adressé à l'Assemblée Nationale* (Paris, 1790).

CARON, J.-F.-C., *Réflexions sur l'exercice de la médecine* (Paris, 1804).

——, *Projet de règlement sur l'art de guérir* (Paris, 1801).

CHAMBON DE MONTAUX, *Moyens de rendre les hôpitaux utiles et de perfectionner la médecine* (Paris, 1787).

COLON DE DIVOL, *Réclamations des malades de Bicêtre* (Paris, 1790).

COQUEAU, C.-P., *Essai sur l'établissement des hôpitaux dans les grandes villes* (Paris, 1787).

DAUNOU, P.-C., *Rapports sur les Ecoles spéciales* (Paris, Year V).

DEMANGEON, J.-B., *Tableau d'un triple établissement réuni en un seul hospice à Copenhague* (Paris, Year VII).

——, *Des moyens de perfectionner la médecine* (Paris, 1804).

DESMONCEAUX, A., *De la bienfaisance nationale* (Paris, 1787).

DUCHANOY, *Projet d'organisation médicale* (n.p., n.d.).

DU LAURENS, J., *Moyens de rendre les hôpitaux utiles et de perfectionner les médecins* (Paris, 1787).

DUPONT DE NEMOURS, P., *Idées sur les secours à donner aux pauvres malades dans une grande ville* (Paris, 1786).

EHRMANN, J.-F., *Opinion sur le projet de Vitet, 14 germinal an VI* (Paris, Year VI).

Essai sur la réformation de la société dite de médecine (Paris, Year VI).

Etat actuel de l'Ecole de Santé (Paris, Year VI).

FOURCROY, A. F., *Discours sur le projet de loi relatif à l'exercice de la médecine, 19 ventôse an XI* (Paris, Year XI).

——, *Exposé des motifs du projet de loi relatif à l'exercice de la médecine* (Paris, n.d.).

——, *Rapport sur les Ecoles de Médecine, frimaire an III* (Paris, Year III).

——, *Rapport sur l'enseignement libre des sciences et des arts* (Paris, Year II).

FOUROT, *Essai sur les concours en médecine* (Paris, 1786).

GALLOT, J.-G., *Vues générales sur la restauration de l'art de guérir* (Paris, 1790).

GÉRAUD, M., *Projet de décret à rendre sur l'organisation civile des médecins* (Paris, 1791).

GUILLAUME, J., *Procès-verbaux du Comité d'Instruction publique* (Paris, 1899).

GUILLEMARDET, F.-P., *Opinion sur les Ecoles spéciales de Santé, 14 germinal an VI* (Paris, Year VI).

IMBERT, J., *Le droit hospitalier de la Révolution et de l'Empire* (Paris, 1954).

Instituta facultatis medicae Vidobonensis, curante Ant. Storck (Vienna, 1775).

JADELOT, N., *Adresse à Nos Seigneurs de l'Assemblée Nationale sur la nécessité et les moyens de perfectionner l'enseignement de la médecine* (Nancy, 1790).

LEFÈVRE, J., *Opinion sur le projet de Vitet, 16 germinal an VI* (Paris, Year VI).

LESPAGNOL, N.-L., *Projet d'établir trois médecins par district pour le soulagement des gens de la campagne* (Charleville, 1790).

MARQUAIS, J.-Th., *Rapport au Roi sur l'état actuel de la médecine en France* (Paris, 1814).

MENURET, J.-J., *Essai sur les moyens de former de bons médecins* (Paris, 1791).

Motif de la réclamation de la Faculté de Médecine de Paris contre l'établissement de la Société royale de Médecine (n.p., n.d.; the author is VACHER DE LA FEUTRIE).

Observations sur les moyens de perfectionner l'enseignement de la médecine en France (Montpellier, Year V).

PASTORET, C.-E., *Rapport sur un mode provisoire d'examen pour les officiers de Santé (19 thermidor an V)* (Paris, Year V).

PETIT, M. A., *Projet de réforme sur l'exercise de la médecine en France* (Paris, 1791).

———, *Sur la meilleure manière de construire un hôpital* (Paris, 1774).

Plan de travail présenté à la Société de Médecine de Paris (Paris, Year V).

Plan général d'enseignement dans l'Ecole de Santé de Paris (Paris, Year III).

PORCHER, G.-C., *Opinion sur la résolution du 19 fructidor an V, 16 vendémiaire an VI* (Paris, Year VI).

Précis historique de l'établissement de la Société royale de Médecine (n.p., n.d.).

PRIEUR DE LA CÔTE-D'OR, C.-A., *Motion relative aux Ecoles de Santé* (Paris, Year VI).

Programme de la Société royale de Médecine sur les cliniques (Paris, 1792).

Programme des cours d'enseignement dans l'Ecole de Santé de Montpellier (Paris, Year III).

PRUNELLE, Cl.-V., *Des Ecoles de Médecine, de leurs connexions et de leur methodologie* (Paris, 1816).

Recueil de discours prononcés à la Faculté de Montpellier (Montpellier, 1820).

RÉGNAULT, J.-B., *Considérations sur l'état de la médecine en France depuis la Révolution jusqu'à nos jours* (Paris, 1819).

RETZ, N., *Exposé succinct à l'Assemblée Nationale sur les Facultés et Sociétés de Médecine* (Paris, 1790).

ROYER, P.-F., *Bienfaisance médicale et projet financier* (Provins, Year IX).

———, *Bienfaisance médicale rurale* (Troyes, 1814).

SABAROT DE L'AVERNIÈRE, *Vue de législation médicale adressée aux Etats généraux* (n.p., 1789).

TISSOT, S.-A.-D., *Essai sur les moyens de perfectionner les études de médecine* (Lausanne, 1785).

VICQ D'AZYR, F., *Œuvres* (6 vols., Paris, 1805).

VITET, L., *Rapport sur les Ecoles de Santé, 17 ventôse an VI* (Paris, Year VI).

WÜRTZ, *Mémoire sur l'établissement des Ecoles de Médecine pratique* (Paris, 1784).

IV. METHODS

AMARD, L.-V.-F., *Association intellectuelle* (2 vols., Paris, 1821).

AMOREUX, P.-J., *Essai sur la médecine des Arabes* (Montpellier, 1805).

AUDIBERT-CAILLE, J.-M., *Mémoire sur l'utilité de l'analogie en médecine* (Montpellier, 1814).

AUENBRUGGER, *Nouvelle méthode pour reconnaître les maladies internes* (Fr. trans. in ROZIÈRE DE LA CHASSAIGNE, *Manuel des pulmoniques*, Paris, 1763).

BEULLAC, J.-P., *Nouveau guide de l'étudiant en médecine* (Paris, 1824).

BORDEU, Th., *Recherches sur le pouls* (4 vols., Paris, 1779–1786).

BOUILLAUD, J., *Dissertation sur les généralités de la clinique* (Paris, 1831).

BROUSSONNET, J.-L.-V., *Tableau élémentaire de séméiotique* (Montpellier, Year VI).

BRULLEY, C.-A., *Essai sur l'art de conjecturer en médecine* (Paris, Year X).

BRUTÉ, S.-G.-G., *Essai sur l'histoire et les avantages des institutions cliniques* (Paris, 1803).

CHOMEL, J.-B.-L., *Essai historique sur la médecine en France* (Paris, 1762).

CLOS DE SORÈZE, J.-A., *De l'analyse en médecine* (Montpellier, Year V).

CORVISART, J.-N., *Essai sur les maladies et lésions du cœur et des gros vaisseaux* (Paris, 1806).

DARDONVILLE, H., *Réflexions pratiques sur les dangers des systèmes en médecine* (Paris, 1818).

DEMORCY-DELETTRE, J.-B.-E., *Essai sur l'analyse appliquée au perfectionnement de la médecine* (Paris, 1818).

DOUBLE, F.-J., *Séméiologie générale ou Traité des signes et de leur valeur dans les maladies* (3 vols., Paris, 1811–1822).

DUVIVIER, P.-H., *De la médecine considérée comme science et comme art* (Paris, 1826).

ESSYG, *Traité du diagnostic médical* (Fr. trans., Paris, Year XII).

FABRE, *Recherche des vrais principes de l'art de guérir* (Paris, 1790).

FORDYCE, G., *Essai d'un nouveau plan d'observations médicales* (Fr. trans., Paris, 1811).

FOUQUET, H., *Discours sur la clinique* (Montpellier, Year XI).

FRANK, J.-P., *Ratio instituti clinici Vicinensis* (Vienna, 1797).

GILBERT, N.-P., *Les théories médicales modernes comparées entre elles* (Paris, Year VII).

GIRBAL, A., *Essai sur l'esprit de la clinique médicale de Montpellier* (Montpellier, 1857).

GOULIN, J., *Mémoires sur l'histoire de la médecine* (Paris, 1779).

HÉLIAN, M., *Dictionnaire de diagnostic ou l'art de connaître les maladies* (Paris, 1771).

HILDENBRAND, J., *Médecine pratique* (Fr. trans., Paris, 1824, 2 vols.).

LANDRÉ-BEAUVAIS, A.-J., *Séméiotique ou traité des signes des maladies* (Paris, 1810).

LEROUX, J.-J., *Cours sur les généralités de la médecine* (Paris, 1818).

———, *Ecole de Médecine. Clinique interne* (Paris, 1809).

LORDAT, J., *Conseils sur la manière d'étudier la physiologie de l'homme* (Montpellier, 1813).

———, *Perpétuité de la médecine* (Montpellier, 1837).

MAHON, P.-A.-O., *Histoire de la médecine clinique* (Paris, Year XII).

MARTINET, L., *Manuel de clinique* (Paris, 1825).

MAYGRIER, J.-P., *Guide de l'étudiant en médecine* (Paris, 1807).

MENURET, J.-J., *Traité du pouls* (Paris, 1798).

MOSCATI, P., *De l'emploi des systèmes dans la médecine pratique* (Strasbourg, Year III).

PETIT, M.-A., *Collection d'observations cliniques* (Lyons, 1815).

PINEL, Ph., *Médecine clinique* (Paris, 1802).

PIORRY, P. A., *Tableau indiquant la manière d'examiner et d'interroger le malade* (Paris, 1832).

ROSTAN, L., *Traité élémentaire de diagnostic, de pronostic, d'indications thérapeutiques* (6 vols., Paris, 1826).

ROUCHER-DERATTE, Cl., *Leçons sur l'art d'observer* (Paris, 1807).

SELLE, Ch.-G., *Médecine clinique* (Fr. trans., Montpellier, 1787).

———, *Introduction à l'étude de la nature et de la médecine* (trad., Montpellier, Year III).

SÉNEBIER, J., *Essai sur l'art d'observer et de faire des expériences* (3 vols., 1802).

THIERY, F., *La médecine expérimentale* (Paris, 1755).

VAIDY, J.-V.-F., *Plan d'études médicales à l'usage des aspirants* (Paris, 1816).

ZIMMERMANN, G., *Traité de l'expérience en médecine* (Fr. trans., Paris, 1774, 3 vols.).

V. MORBID ANATOMY

BAILLIE, M., *Anatomie pathologique des organes les plus importants du corps humain* (Fr. trans., Paris, 1815).

BAYLE, G.-L., *Recherches sur la phtisie pulmonaire* (Paris, 1810).

BICHAT, X., *Anatomie générale appliquée à la physiologie et à médecine* (3 vols., Paris, 1801).

———, *Anatomie pathologique* (Paris, 1825).

———, *Recherches physiologiques sur la vie et la mort* (Paris, Year VIII).

———, *Traité des membranes* (Paris, 1807).

BONET, Th., *Sepulchretum* (3 vols., Lyons, 1700).

BRESCHET, G., *Répertoire général d'anatomie et de physiologie pathologiques* (6 vols., Paris, 1826–1828).

CAILLIOT, L., *Eléments de pathologie et de physiologie pathologique* (2 vols., Paris, 1819).

CHOMEL, A.-F., *Eléments de pathologie générale* (Paris, 1817).

CRUVEILHIER, J., *Essai sur l'anatomie pathologique en général* (2 vols., Paris, 1816).

DEZEIMERIS, J.-E., *Aperçu rapide des découvertes en anatomie pathologique* (Paris, 1830).

GUILLAUME, A., *De l'influence de l'anatomie pathologique sur les progrès de la médecine* (Dôle, 1834).

LAËNNEC, R., *Traité de l'auscultation médiate* (2 vols., Paris, 1819).

———, *Traité inédit de l'anatomie pathologique* (Paris, 1884).

LALLEMAND, F., *Recherches anatomo-pathologiques sur l'encéphale et ses dépendances* (2 vols., Paris, 1820).

MORGAGNI, J.-B., *De sedibus et causis morborum* (Venice, 1761).

PORTAL, A., *Cours d'anatomie médicale* (5 vols., Paris, Year XII).

Prost, P.-A., *La médecine éclairée par l'observation et l'ouverture des corps* (2 vols., Paris, Year XII).

Rayer, P., *Sommaire d'une histoire abrégée de l'anatomie pathologique* (Paris, 1818).

Ribes, Fr., *De l'anatomie pathologique considérée dans ses vrais rapports avec la science des maladies* (2 vols., Paris, 1828-1834).

Richerand, B.-A., *Histoire des progrès récents de la chirurgie* (Paris, 1825).

Saucerotte, C., *De l'influence de l'anatomie pathologique sur les progrès de la médecine* (Paris, 1834).

Tacheron, C.-F., *Recherches anatomo-pathologiques sur la médecine pratique* (3 vols., Paris, 1823).

VI. FEVERS

Barbier, J.-B.-G., *Réflexions sur les fièvres* (Paris, 1822).

Boisseau, F.-G., *Pyrétologie physiologique* (Paris, 1823).

Bompart, A., *Description de la fièvre adynamique* (Paris, 1815).

Bouillaud, J., *Traité clinique ou expérimental des fièvres dites essentielles* (Paris, 1830).

Broussais, F.-J.-V., *Catéchisme de médecine physiologique* (Paris, 1824).

————, *Examen des doctrines médicales* (Paris, 1821).

————, *Histoire des phlegmasies ou inflammations chroniques* (Paris, 1808, 2 vols.).

————, *Leçons sur la phlegmasie gastrique* (Paris, 1819).

————, *Mémoire sur l'influence que les travaux des médecins physiologistes ont exercée sur l'état de la médecine* (Paris, 1832).

————, *Traité de physiologie appliquée à la pathologie* (2 vols., 1822-1823).

Caffin, J.-F., *Quelques mots de réponse à un ouvrage de M. Broussais* (Paris, 1818).

Castel, L., *Réfutation de la nouvelle doctrine médicale de M. le Dr. Broussais* (Paris, 1824).

Chambon de Montaux, *Traité de la fièvre maligne simple et des fièvres compliquées de malignité* (4 vols., Paris, 1787).

Chauffard, H., *Traité sur les fièvres prétendues essentielles* (Paris, 1825).

Chomel, A. F., *De l'existence des fièvres* (Paris, 1820).

————, *Des fièvres et des maladies pestilentielles* (Paris, 1821).

Collineau, J.-C., *Peut-on mettre en doute l'existence des fièvres essentielles* (Paris, 1823).

Dagoumer, Th., *Précis historique de la fièvre* (Paris, 1831).

Dardonville, H., *Mémoire sur les fièvres* (Paris, 1821).

Ducamp, Th., *Réflexions critiques sur les écrits de M. Chomel* (Paris, 1821).

Fodéra, M., *Histoire de quelques doctrines médicales comparées à celles de M. Broussais* (Paris, 1818).

Fournier, M., *Observations sur les fièvres putrides et malignes* (Dijon, 1775).

Gérard, M., *Peut-on mettre en doute l'existence des fièvres essentielles?* (Paris, 1823).

Giannini, *De la nature des fièvres* (Fr. trans., Paris, 1808).

Giraudy, Ch., *De la fièvre* (Paris, 1821).

GRIMAUD, M. de, *Cours complet ou Traité des fièvres* (3 vols., Montpellier, 1791).

HERNANDEZ, J.-F., *Essai sur le typhus* (Paris, 1816).

HOFFMANN, F., *Traité des fièvres* (Fr. trans., Paris, 1746).

HUFELAND, C.-W., *Observations sur les fièvres nerveuses* (Fr. trans., Berlin, 1807).

HUXHAM, J., *Essai sur les différentes espèces de fièvres* (Fr. trans., Paris, 1746).

LARROQUE, J.-B. de, *Observations cliniques opposées à l'examen de la nouvelle doctrine* (Paris, 1818).

LEROUX, F.-M., *Opposition aux erreurs de la science médicale* (Paris, 1817).

LESAGE, L.-A., *Danger et absurdité de la doctrine physiologique* (Paris, 1823).

MONFALCON, J.-B., *Essai pour servir à l'histoire des fièvres adynamiques* (Lyons, 1823).

MONGELLAZ, P.-J., *Essai sur les irritations intermittentes* (2 vols., Paris, 1821).

PASCAL, Ph., *Tableau synoptique du diagnostic des fièvres essentielles* (Paris, 1818).

PETIT, M.-A., *Traité de la fièvre entéro-mésentérique* (Paris, 1813).

PETIT-RADEL, Ph., *Pyrétologie médicale* (Paris, 1812).

QUITARD-PIORRY, H.-H., *Traité sur la non-existence des fièvres essentielles* (Paris, 1830).

ROCHE, L.-Ch., *Réfutation des objections faites à la nouvelle doctrine des fièvres* (Paris, 1821).

ROEDERER and WAGLER, *Tractatus de morbo mucoso* (Göttingen, 1783).

ROUX, G., *Traité des fièvres adynamiques* (Paris, 1812).

SELLE, Ch.-G., *Eléments de pyrétologie méthodique* (Fr. trans., Lyons, Year IX).

STOLL, M., *Aphorismes sur la connaissance et la curation des fièvres* (Fr. trans., Paris, Year V).

TISSOT, S.-A.-D., *Dissertation sur les fièvres bilieuses* (Fr. trans., Paris, Year VIII).

Index

Numbers in brackets are references to notes, indicating that
although the person in question is not named in the text at that
point, the number in brackets will be found at the end of quoted
or paraphrased material derived from his works; author and title
are cited in the corresponding note at the end of the chapter in
which the reference appears.

Fourcroy, Comte Antoine-François
  de, 49, 50 [32], 69–72, 74, 76,
  81 [53]
Franck, J.-P., 57, 181
Freud, Sigmund, 198
Frier, François, 8 [13]

Gallot, J.-G., 47
Géraud, Mathieu, 29 [23], 30
Géricault, Jean-Louis-André-Théo-
  dore, 171
Gilibert, Jean-Immanuel, 4
Girbal, A., 68 [18]
Göttingen, clinic at, 57
Goya y Lucientes, Francisco José
  de, 171
Grenoble, medical society at, 74
Grimaud, Jean-Charles-Marquerite
  de, 180
Guardia, 198
Guilbert, 57
Guillaume, J., 66 [10]
Guindant, Toussaint, 9 [14]

Haen, Anton de, 57
Haller, Albrecht von, 181, 188, 192
Haslam, John, 11 [16]
Hautesierck, François-Marie-
  Claude Richard de, 28–9
Hildenbrand, Johann Valentin, 57
Hippocrates, 56, 107
Hölderlin, Friedrich, 197, 198
Hôpital de l'École, 70
Horne, Jan van, 130
Hospice de l'Humanité, 70
Hospice de l'Unité, 70
Hôtel-Dieu (Paris), 61, 125
Hunter, John, 125, 134, 141
Huxham, John, 179

Jackson, 198
Jadelot, N., 20 [40], 46 [22]

Kant, Immanuel, xv
Knips, 57

La Boe, François de, 57
Lacassaigne, 57
Laënnec, René-Théophile-Hya-
  cinthe, 132–3, 135, 143–4, 151,
  157, 160, 163, 164 [39], 165, 167,
  168, 169 [53], 170 [54], 176,
  177, 178
Lafisse, 73
Lallemand, Claude-François, xii-
  xiii, 152 [13], 157 [26], 191
Lamartine, Alphonse-Marie-Louis
  de Prat de, 171
Landré-Beauvais, A.-J., 93 [10, 11],
  94 [13, 14]
Lanthenas, F., 33 [33, 34, 35], 34
  [36], 35
Larrey, J.-D., 74
Lasson, 27
Lavoisier, Antoine-Laurent, 131
Lebon, 43
Le Brun, L.-S.-D., 23 [3], 24 [8],
  25 [10, 12], 26 [14]
Lepecq de La Clôture, 23 [4]
Le Pelletier de Saint-Fargeau,
  Louis-Michel, 50
Leveillé, 73
Leyden, clinical school at, 57
Lieutaud, Joseph, 126, 130
Lioult, P.-J., 65 [4]
London, teaching hospital at, 57
Lyons, medical society at, 74

Magendie, François, 145
Mahon, P.-A.-O., 56 [6, 11, 12]
Malebranche, Nicolas de, xiii
Maret, M., 33 [32]
Martin, 156, 177 [12]
Maygrier, J.-P., 110 [9]
Meckel, Johann Friedrich, xii, 12–
  13
Menuret, J.-J., 19, 23 [6], 24 [9],
  29 [21], 32 [31], 84 [59], 161
Michelet, Jules, 125
Mondor, Henri, 178
Montpellier, medical school at, 75–
  76

THE ORDER OF THINGS
An Archaeology of Human Sciences

With vast erudition, Foucault cuts across disciplines and reaches back into seventeenth century to show how classical systems of knowledge, which linked all of nature within a great chain of being and analogies between the stars in the heavens and the features in a human face, gave way to the modern sciences of biology, philology, and political economy. The result is nothing less than an unearthing of old patterns of meaning and a revelation of the shocking arbitrariness of our received truths. In the work that established him as the most important French thinker since Sartre, Michel Foucault offers startling evidence that "man"—man as a subject of scientific knowledge—is at best a recent invention, the result of a fundamental mutation in our culture.

Philosophy/History

MADNESS AND CIVILIZATION
A History of Insanity in the Age of Reason

Michel Foucault examines the archeology of madness in the West from 1500 to 1800—from the late Middle Ages, when insanity was still considered part of everyday life and fools and lunatics walked the streets freely, to the time when such people began to be considered a threat, asylums were first built, and walls were erected between the "insane" and the rest of humanity.

History/Psychology

Not one of Michel Foucault's books offers a satisfactory introduction to the entire complex body of his work. *The Foucault Reader* was commissioned precisely to serve that purpose. The *Reader* contains selections from each area of Foucault's work as well as a wealth of previously unpublished writings, including important material written especially for this volume, the preface to the long-awaited second volume of *The History of Sexuality*, and interviews with Foucault himself, in which he discussed his philosophy firsthand, with unprecedented candor. This philosophy comprises an astonishing intellectual enterprise: a minute and ongoing investigation of the nature of power in society. Foucault's analyses of this power as it manifests itself in society, schools, hospitals, factories, homes, families, and other forms of organized society are brought together in *The Foucault Reader* to create an overview of this theme and of the broad social and political vision that underlies it.

Philosophy/History

ALSO AVAILABLE
The Archaelogy of Knowledge
Discipline and Punish
Herculine Barbin
The History of Sexuality
The History of Sexuality, Vol. Two
The History of Sexuality, Vol. Three
Power/Knowledge

VINTAGE BOOKS
Available wherever books are sold.
www.vintagebooks.com